Water Security in a New World

Series editor
Zafar Adeel, Pacific Water Research Centre, Simon Fraser University, Burnaby, BC, Canada

More information about this series at http://www.springer.com/series/13406

David Devlaeminck • Zafar Adeel
Robert Sandford
Editors

The Human Face of Water Security

Springer

Editors
David Devlaeminck
China International Water Law (CIWL), School of Law
Xiamen University
Xiamen, Fujian, China

Zafar Adeel
Pacific Water Research Centre
Simon Fraser University
Burnaby, BC, Canada

Robert Sandford
Hamilton, ON, Canada

ISSN 2367-4008 ISSN 2367-4016 (electronic)
Water Security in a New World
ISBN 978-3-319-50160-4 ISBN 978-3-319-50161-1 (eBook)
DOI 10.1007/978-3-319-50161-1

Library of Congress Control Number: 2017935378

Printed on acid-free paper

This Springer imprint is published by Springer Nature
The registered company is Springer International Publishing AG
The registered company address is: Gewerbestrasse 11, 6330 Cham, Switzerland

Contents

Contributors

Zafar Adeel Pacific Water Research Centre, Simon Fraser University, Burnaby, BC, Canada

Patricia Avila-Garcia Research Institute in Ecosystems and Sustainability (UNAM-IIES), National Autonomous University of Mexico, Mexico City, Mexico

Alice Bouman-Dentener DiploriA Sustainable Development Solutions, The Hague, The Netherlands

Women for Water Partnership, The Hague, The Netherlands

Sarah Dickin Stockholm Environment Institute, Stockholm, Sweden

Sarah E. Dickson Department of Civil Engineering, McMaster University, Hamilton, ON, Canada

Luca Di Mario Asian Development Bank, Manila, Philippines

Jennifer Fresque-Baxter Department of Environment and Natural Resources, Government of the Northwest Territories, Yellowknife, NT, Canada

Erin Kelly Department of Environment and Natural Resources, Government of the Northwest Territories, Yellowknife, NT, Canada

Kyungmee Kim Department of Peace and Conflict Research, Uppsala University, Uppsala, Sweden

Research School for International Water Cooperation, UNESCO Category II Centre on International Water Cooperation, Uppsala, Sweden

Owen McIntyre School of Law, University College Cork, Cork, Ireland

Robert Sandford United Nations University – Institute for Water, Environment and Health, Canada, Hamilton, ON, Canada

Corinne J. Schuster-Wallace Independent Water-Health Expert, McMaster University, Hamilton, ON, Canada

Ashok Swain Department of Peace and Conflict Research, Uppsala University, Uppsala, Sweden

Research School for International Water Cooperation, UNESCO Category II Centre on International Water Cooperation, Uppsala, Sweden

Susan Watt School of Social Work, McMaster University, Hamilton, ON, Canada

United Nations University Institute for Water, Environment and Health, Hamilton, ON, Canada

List of Abbreviations

AAROM	Aboriginal Aquatic Resources and Oceans Management
ASC	Aboriginal Steering Committee
BOT	Build-operate-transfer
BOOT	Build-own-operate-transfer
BLT	Build-lease-transfer
BWMA	Bilateral Water Management Agreement
CASA 1000	1000 Electricity Transmission and Trade Project for Central and South Asia
CBD	Convention on Biological Diversity
CBM	Community-Based Water Quality Monitoring Program
CESCR	Committee on Economic, Social and Cultural Rights
CPEC	China-Pakistan Economic Corridor
DALYs	Disability-adjusted life years
EIA	Environmental Impact Assessment
ENGO	Environmental nongovernmental organization
ENR-GNWT	Department of Environment and Natural Resources–Government of the Northwest Territories
FAO	Food and Agriculture Organization
FDI	Foreign direct investment
FSDS	Federal Sustainable Development Strategy
GDP	Gross domestic product
GLAAS	Global Analysis and Assessment of Sanitation and Drinking Water
GNI	Gross national income
GNWT	Government of the Northwest Territories
GWP	Global Water Partnership
HIA	Health impact assessment
HIC	High-income country
ICCPR	International Covenant on Civil and Political Rights
ICESCR	International Covenant on Economic, Social and Cultural Rights
ICJ	International Court of Justice
IDP	Internally displaced persons

IFAD	International Fund for Agricultural Development
ILC	International Law Commission
IPCC	Intergovernmental Panel on Climate Change
ISIS	Islamic State in Iraq and Syria
IWRM	Integrated water resources management
KAP	Knowledge, attitudes, and practices
KWDT	Katosi Women Development Trust
LHWP	Lesotho Highland Water Project
LMIC	Low- and middle-income countries
MDG	Millennium Development Goals
MRB	Mackenzie River Basin
MRBB	Mackenzie River Basin Board
NDC	Nationally determined contributions
NGO	Non-governmental organization
NWT	Northwest Territories (Canada)
ODA	Official development aid
OECD	Organisation for Economic Co-operation and Development
OECD-DAC	Organisation for Economic Co-operation and Development–Development Assistance Committee
PCIJ	Permanent Court of International Justice
RADWQ	Rapid assessment of drinking water-quality
RBA	Human rights-based approach
RBO	River Basin Organisation
RRM	Rural, remote, and marginalized communities
RWH	Rainwater harvesting
SDG	Sustainable Development Goal
SIA	Social impact assessment
SRDP	Slave River and Delta Partnership
SWAT	Self-Water Assessment Tool
SWEEP	Slave Watershed Environmental Effects Program
TAPI	Turkmenistan-Afghanistan-Pakistan-India Pipeline
TWG	Tegemeo Women Group
UN	United Nations
UNCED	United Nations Conference on Environment and Development
UNCESCR	United Nations Committee on Economic, Social and Cultural Rights
UNCLOS	United Nations Convention on the Law of the Sea
UNECE	United Nations Economic Commission of Europe
UNEP	United Nations Environment Programme
UNFCCC	United Nations Framework Convention on Climate Change

UNU-INWEH	United Nations University Institute for Water, Environment and Health
UNWC	United Nations Watercourses Convention
UPA	Urban and peri-urban agriculture
USAID	United States Agency for International Development
UWASNET	Uganda Water and Sanitation Network
VIP	Ventilated improved pit latrine
WaSH	Water, sanitation and hygiene
WHO	World Health Organization
WSP	Water safety plan
WOFAN	Women Farmers Advancement Network of Nigeria

Boxes, Figures, and Tables

Boxes

Figures

Tables

Chapter 1
The Human Face of Water Insecurity

Robert Sandford

Abstract The timely availability of fresh water has for decades been recognized as a global concern. In addition to matters of availability and quality we now recognize that the world will soon be redefined by changing precipitation patterns associated with an increase in the mean temperature of our planet's atmosphere. This will result in droughts in some places becoming deeper and more persistent making human presence in some parts of the world impossible to sustain. While it did not receive the same attention in the media, the announcement of UN's 2030 *Transforming Our World* global sustainable development agenda was at least as important as the climate negotiations held in Paris 2 months later if only because it deals with damage we are doing to other elements of the Earth system that are exacerbating and being exacerbated by climate change. The goals in the agenda of improving the management of water globally and acting on climate change need to be elevated to special importance because success cannot be achieved in addressing other critical global sustainable development challenges, which include huge challenges such as eliminating poverty and hunger and bringing about peace and stability, unless we manage water more effectively, a goal that can only be achieved by stabilizing the composition of the Earth's atmosphere. This chapter argues that if we do not make water security and water-related climate stability a global imperative at the national and sub-national level, the result will be greater regional tension, conflict and involuntary migration related in large measure to water insecurity.

Keywords Water security • Climate stability • Conflict • Migration

1.1 Water Security and Insecurity

In general, water security is usually taken to mean having and being able to reliably provide adequate water of the right quality where and when you need it for all purposes especially agriculture but also for purposes related to sustainable natural

---

R. Sandford (✉)
United Nations University – Institute for Water, Environment and Health,
Hamilton, ON, Canada
e-mail: sandford@telusplanet.net; robert.w.sandford@gmail.com

D. Devlaeminck et al. (eds.), *The Human Face of Water Security*, Water Security in a New World, DOI 10.1007/978-3-319-50161-1_1

bio-diversity-based Earth system function. It also means ensuring that your use and management of water in the region in which you live does not significantly affect the water security of regions up or downstream from you now or in the future.

Over the last decade water security has also come to mean being able to achieve these goals not just in the face of growing populations but also in the face of new circumstances created by the acceleration of the global hydrological cycle.

Water security now means managing not just the water you once thought was reliably available to you; but also managing water in ever greater extremes of abundance and scarcity than we have experienced in the past.

What we now realize is that water security and climate security are inseparable; one is implicit in the other. It could even be said they are the same thing. It is widely held that water and climate security are critical elements of sustainability. Without stable water and climate regimes sustainability will forever remain a moving target.

Flood and drought resilience are very much elements of the larger water security ideal in that they are tied to human health and well-being as well as social, economic and political stability especially in parts of the world that are simultaneously experiencing rapid population growth and changes in hydro-meteorological regimes. The failure to assure flood and drought resilience is increasingly seen as contributing to national and regional water in-security that can have devastating local effects which spill out of and affect surrounding regions and may even affect the rest of the world. This is a relatively new development in the politics of hydro-climatic change which is now, because of social media, assuming a very human face.

On Wednesday, September, 2nd, 2015, the body of a 3-year old boy, a Syrian refugee named Aylan Kurdi, later identified Alan Kurdi, washed up on a beach in Turkey. While the boy's face was only partially exposed in the now famous photographs taken of him, and though he drowned in water, Alan Kurdi put a child's face on the global threat of water insecurity. Here was a little boy dressed in a red shirt; blue pants and running shoes – a child that could have been anyone's child anywhere – alone and face down in the incoming waves. The images, posted by the Turkish photographer Nilüfer Demir and British photo-journalist Andy Worthington, were instantly communicated around the world through social media causing immediate international outrage over the world's failure to understand the serious nature of what was happening in Syria and to act upon its humanitarian responsibility toward refugees. While some disagreed that Alan Kurdi could be characterized as a true "climate refugee" in that the conflict in Syria could have broken out at any time irrespective of drought, the fact remains that deep and persistent drought did play a role in intensifying the Syrian refugee crisis.

In the 5 years between 2006 and 2011, over half of Syria was engulfed in the worst drought in recorded history. It was held by many climate scientists that the drought was deeper and more persistent than could be explained by established patterns of regional natural climate variability. As the drought worsened, nearly 85% of livestock died and crops upon which locals relied for food and for export withered away in the dry fields. When the government of President Bashar al-Assad responded by granting groundwater well rights along political and ideological lines, many farmers were forced to drill their own illegal wells. As the drought intensified there were demonstrations demanding government action. Protesters were soon being arrested and tortured.

As the drought lingered on, nearly a million farmers were forced to abandon their farms and move to already crowded cities like Daraa, exacerbating already extreme water shortages in those cities. In these cities there were few jobs for those who crowded in from the countryside. Even though most international analysts had deemed Syria immune to what became known around the world as the Arab Spring, angry Syrians were emboldened by what they saw happen on television in Tunis and Cairo. A movement emerged with the stated purpose of toppling the Syrian political regime. The protesters, however, under-estimated Bashar al-Assad's ruthlessness. Further arrests and torture followed. The revolution became violent. Soon other nations and regional extremist movements, each with its own regional interests became involved exacerbating the conflict. At the time of this writing the conflict had yet to be resolved. In the meantime, the surrounding regions, including Europe, are facing the largest forced migration of people since World War II.

Can it be legitimately stated that climate change impacts on water security played a role in the Syrian conflict? There can be no questioning the fact that loss of livelihood by 1.5 million people due to drought put a great deal of stress on Syria's social and economic fabric. Nor is there any question that massive population displacement from rural areas into Syrian cities contributed to social unrest and fueled the revolutionary movement that intensified the war. While it is true as international water expert Peter Gleick noted in his paper "Water, Drought, Climate and Change in Syria" that the civil war that began in March of 2011 was sparked by the complex interaction of a number of social and political factors both nationally and regionally, there has clearly been an increase in incidents of water-related violence at sub-national levels around the world. Syria is one of these places. Many of these conflicts are attributable to the role water security plays in maintaining national and regional economic, social and political stability.

1.2 The World Had Been Warned

The world has been warned many times that the impacts of water insecurity could radiate outward from the water-scarce regions of the Middle East and North Africa in ways that would have impacts on Europe and the rest of the world.

Established in 1983, the InterAction Council is an independent international organization that aims to mobilize the experience, energy and international contacts of a group of statesmen who have held the highest political office in their own countries. Its members jointly develop recommendations on — and practical solutions for — political, economic and social problems confronting all of humanity. The council is unique in bringing together on a regular basis more than 30 former heads of state or government who work together to foster international co-operation in three principal areas — peace and security, revitalization of the world economy and the nexus between development, population and environment and universal ethics. The members of the InterAction Council select specific issues from within these broad areas and then develop proposals for action. They then communicate these proposals directly to government leaders and other national decision-makers, heads of international organizations and influential individuals throughout the world. If there could be said to be a global senate then this might be it.

The final communiqué of the 30th Annual Plenary Meeting of the Interaction Council held in China in 2012 reported (InterAction Council 2012) that, as a result of humanity's over-exploitation and pollution of water resources, there is now a growing global water crisis. If left unaddressed, the Council noted, water scarcity, and the deteriorating water environment will undermine human health, hinder economic development and in some places will even affect national and regional stability.

The demand for water continues to grow in tandem with rapidly increasing human populations and accompanying economic and social development. It must be realized that, despite this growth, the world cannot continue to divert water indefinitely for consumptive purposes. Enough water must remain in the environment to sustain biodiversity-based planetary life-support system functions. Aquatic ecosystems must be protected to safeguard natural processes of water purification, prevent floods, and moderate the effects of drought.

While the 30th Annual Plenary was held in China, which has its own serious water problems, it was noted at that meeting that the challenges associated with water security were particularly serious in the Middle East and North Africa. The flows of the Jordan River, for example, have diminished by 95% due to dams and diversions affecting not only Jordan but all the countries that rely upon its waters. The political tensions and distrust that have characterized the region in the past must be resolved before cooperation can lead to more efficient water management. But no solution to the water scarcity problems in the Middle East and North Africa, the InterAction Council noted, will be possible without on-going dialogue between decision-makers and water experts.

The InterAction Council also recommended that conflict over water can be avoided by adopting the principles of basin-scale management and cooperation. In conclusion, the InterAction Council recommended enhanced public education campaigns aimed at making the global public more aware of water issues. A further examination of water scarcity issues in the Middle East and North Africa has confirmed the validity and urgency of these recommendations.

The Middle East and North Africa region faces serious food security issues and some of the most daunting climate change challenges of any region in the world. Because of the besieged circumstances of many governments and the social and political instability of the past several decades, water and water-related food and energy security issues in the region are being addressed in a short-sighted manner that will inevitably lead to conflict as populations grow out of proportion to local food supply and changing climatic conditions reduce available water supplies.

A clear-sighted, long term view of the water scarcity problem is required now not just to lessen the potential for tension between states over water supply and quality issues but to prevent outright collapse of some nations in the Middle East and North Africa region in the near future. That longer view can be provided in part by science.

Food security is a serious concern in many countries in the Middle East and North Africa. The food crisis that began in 2008 has worsened. In 2012 alone, international wheat prices rose by 19%. With each rise in the food price index millions around the world are pushed into poverty. High food prices hit the poor hardest which results in greater unrest. If the state is unable to make up the difference in prices through subsidies; or if food staples of the quantity and quality required are not available in the international market; then unrest grows. When this happens the issue of food security becomes or is absorbed into larger issues of state and regional security.

The Middle East and North Africa have the largest food deficit of any region in the world. Relative food security exists only in the oil-rich countries of the Arabian Peninsula which together constitute only 10% of the region's population. At moderate risk are Iran, Libya and Tunisia. Jordan already faces a serious food security situation, and extremely alarming food security situations exist now in Yemen and Sudan.

Despite already serious water scarcity and a clear understanding of the huge threats to water security in the region, water continues to be badly misused in agriculture. Farmers in many areas are still permitted to abstract water unsustainably and to draw down aquifers in excess of recharge depleting centuries-old aquifers. This kind of mismanagement accelerates environmental degradation through soil erosion, soil and water salination and through water-logging. These are of course global problems but ones that are particularly serious in dry areas like the Middle East and North Africa.

While the outlook for food security in the region is not good, research demonstrates that well-run trans-national research institutions can assist all the nations in the Middle East and North Africa region to simultaneously improve agricultural productivity and practices while at the same time put into relief policy directions that will close the food security gap.

Yield gap has been defined as the amount that actual yields from either irrigated or rain-fed areas fall short of potential yields under optimal management. It is generally held that agricultural productivity gains can be most effectively achieved by reducing the yield gap and increasing relative land availability. Unfortunately, very little capacity exists to increase relative land availability in the Middle East and North Africa region which means that in order to enhance food security, the focus has to be on reducing the yield gap.

Yield gap analysis conducted by researchers in Morocco and Syria demonstrate that the gap between actual and simulated potential yields in research facilities was as high as 82% which suggests a huge potential for increasing wheat production in many parts of the Middle East and North Africa region.

These analyses demonstrate that even without climate change, business as usual is not an option in the Middle East or North Africa. Even with improved agricultural productivity, population growth will put food security even further beyond reach for many countries in the region. Questions remain, however, about whether gains of this order can be achieved at the farm level and if they can be sustained under new climate regimes that drastically reduce water supplies.

Climate change effects are already measureable. The relative change in mean annual temperature and precipitation is higher in the Middle East and North Africa region than anywhere in the world. Temperatures and evaporation rates are expected to rise and precipitation to fall putting even greater pressure on limited water supplies.

It is presently anticipated that the Middle East and North Africa as a region will likely have up to one-quarter less water just as temperatures rise and the population doubles. This situation is projected to be even worse in Jordan and in surrounding countries. It has been predicted that the average water yield in Jordan will decrease by 45–60% due to a 10% decrease in precipitation combined with the evaporative effect of a 2 °C temperature increase. Unfortunately, however, the mean temperature increase is expected to be as high as 4.5 °C by the end of the century with a decrease in precipitation of as much as 25% with combined effects leading to a decrease in water availability of 23% in the upper Jordan catchment. Jordan is not alone in facing this magnitude of hydro-climatic change. It is anticipated that per capita water availability will drop by 35% in Iraq; 38% in Morocco; and 40% in Yemen by 2025. The implications of these changes are far reaching to say the least.

Countries with high reliance on dryland agriculture will be increasingly vulnerable to seasonal changes in climatic patterns. The rural poor will be disproportionately impacted by climate disruption because of their greater dependence on agriculture and their limited capacity to adapt to such changes. The InterAction Council warned that issues of food security could impact political stability in ways that could further threaten the fragile stability of the region. This realization makes building a more effective bridge between science and public policy even more urgent.

Given the current low level of cooperation at the nexus of water, energy and food in the Middle East, the very survival of some countries in the region is threatened by an uncoordinated response to accelerating climate effects on precipitation and evaporation. Unless we are successful in building a more effective bridge between science and public policy not just on a national basis but regionally, some countries may cease to exist at least as we know them today.

Presently the Middle East and North Africa are too absorbed with tensions and uncertainty in their own region to look far into the future. Unfortunately, however, the future as projected presently by science is one that threatens to devastate the region. If current trends persist, growing regional tensions related to water could destabilize the Middle East with impacts that will be felt throughout the rest of the world. Only the international community can prevent that from happening.

1.3 Water in the World We Want

The timely availability of fresh water has for decades been recognized as a global concern. There is not enough water to support our constantly growing numbers and to sustain all the uses to which we want to put this precious resource. In addition to matters of availability and quality we now recognize that the world will soon be redefined by changing precipitation patterns associated with an increase in the mean temperature of our planet's atmosphere. There are going to be winners and losers – places that will remain habitable and places that will not. The geography of human presence on the planet is about to change. Change in that geography is unlikely to occur without conflict. While there are precedents to suggest that outright warfare specifically over water can be avoided, solving the problem of inequitable water supply and reducing the tensions persistent water shortages create will not be easy nor will it be cheap.

Among the many reports published by the UN in the lead-up to the Paris climate conference in 2015, was one published by the United Nations University Institute for Water, Environment and Health which warned that without large new water-related investments many societies worldwide will soon confront rising desperation and conflicts over life's most essential resource. Presenting their report which was entitled *Water in the World We Want: Catalysing National Water-Related Sustainable Development* (Schuster-Wallace and Sandford 2015), at UN Headquarters in New York, officials of UN University and the UN Office for Sustainable Development said unmet water goals threaten many regions of the world and form a barrier to key universally-shared ambitions including stable political systems, greater wealth, and better health for all.

The *Water in the World We Want* report argued that continued stalling, coupled with population growth, economic instability, disrupted climate patterns and other variables, could reverse hard-earned development gains and preclude meaningful levels of development that can be sustained into the future. The report underlined that all current water management challenges will be compounded one way or another by hydro-climatic change, and by increasingly unpredictable weather. The report noted that historical predictability, known as relative hydrological stationarity provided the certainty needed to build houses to withstand winds of a certain speed, snow of a certain weight, and rainfalls of certain intensity and duration, when to plant crops, and to what size to build storm sewers. Because of warming generated by changes in the composition of the Earth's atmosphere, however, the relative stability of the global hydrological cycle has been lost. The consequence is that the management of water in all its forms in the future will involve a great deal more uncertainty than it has in the past.

In a more or less stable hydro-climatic regime, the report observed, you are playing poker with a deck that you know and can bet on risk accordingly. The loss of stationarity is playing poker with a deck in which new cards you have never seen before keep appearing more and more often, ultimately disrupting your hand to such an extent that the game no longer has coherence or meaning. People, unfortunately, do not have the luxury of living without water and when faced with a life or death decision, people tend to do whatever they must to survive. Changes in fundamental hydrology, the report noted, "are likely to cause new kinds of conflict, and it can be expected that both water scarcity and flooding will become major trans-boundary water issues."

The report noted that within 10 years, researchers predict 48 countries — 25% of all nations on Earth with an expected combined population of 2.9 billion — will be classified "water-scarce" (1000–1700 cubic meters of water per capita per year) or "water-stressed" (1000 cubic meters or less). By 2030, overall global demand for freshwater could exceed supply by 40%, with the most acute problems in warmer, low-resource nations with young, fast-growing populations, according to the report.

An estimated 25% of the world's major river basins run dry for part of each year, the report noted, and new conflicts are likely to emerge as more of the world's rivers become further heavily abstracted so that they no longer make it to the sea. Meanwhile, the magnitude of floods in Pakistan and Australia in 2010, and on the Great Plains of North America in 2011 and 2014, "suggests that the destruction of upstream flood protection and the failure to provide adequate downstream flood warning will enter into global conflict formulae in the future." The report cited the rising cost of world flood-related damages: US$53 billion in 2013 and more than US$312 billion since 2004.

Published in the run-up to the UN Member State adoption of universal post-2015 Sustainable Development Goals, the report provided an in-depth analysis of 10 countries to show how ensuring reliable water supply and providing universal sanitation services offers a rapid, cost effective way to achieve sustainable development. The countries included in the study cover the full range of economic and development spectrum: Bangladesh, Bolivia, Canada, Indonesia, Republic of Korea, Pakistan, Singapore, Uganda, Vietnam, and Zambia. Based on the national case studies, the report prescribes country level steps for achieving the global water targets.

The report also noted that the success of global efforts to achieve sustainable development goals with respect to water on the scale required rests in large part on a crackdown on widespread corruption in the water sector, particularly in developing countries. The report noted that in many places corruption is resulting in the hemorrhaging of precious financial resources, siphoning an estimated 30% of funds earmarked for water and sanitation-related improvements.

The *Water in the World We Want* report also noted that the world's water and wastewater infrastructure maintenance and replacement deficit is building at a rate of $200 million per year, with $1 trillion now required in the USA alone. To finance its recommendations, the report says that in addition to plugging the leakage of funds to corruption, $1.9 trillion in subsidies to petroleum, coal and gas industries should be redirected by degrees. The estimated global cost to achieve post-2015 sustainable development goals in water and sanitation development, maintenance and replacement is US $1.25–$2.25 trillion per year for 20 years, a doubling or tripling of current spending translating into 1.8–2.5% of global GDP. The resulting benefits would be commensurately large, however — a minimum of $3.11 trillion per year, not counting health care savings and valuable ecosystem service enhancements.

The report recommended that national governments must make sustainable advancements in water, wastewater, and sanitation management, supported by a dedicated and independent arm's length water agency, a high level policy priority. The report also observed that decisions for managing water at all scales must be evidence informed, accounting for the multiple roles, uses, and demands on water and disposal of human waste and wastewater, as well as the way in which the distribution of water resources is changing, and expected to continue to change over time and space.

Governments and all economic sectors, the report clearly stated, must eradicate corruption through the establishment and implementation of clear and defined anti-corruption protocols, with meaningful consequences when the protocols are breached. Capacity development must be nested within, and form a pillar of institutional reform at all scales within a country, with an emphasis on transferable skills that can be used for sustainable development across all areas and goals. The report also recommended that governments, supported by relevant stakeholders, must commit to timely and transparent monitoring and reporting on sustainable development indicators to monitor progress and hold the global community mutually accountable. There must also be a national commitment to universal access to water and sanitation, linked to waste treatment and management, delivered through nationally coordinated and monitored multi-stakeholder response. The report also noted that the world can no longer ignore the water nature needs for planetary biodiversity based Earth system function and charged national governments with committing to ensuring continued viability and level of provisioning and regulating natural ecosystem functions.

The report recommended that national water governance and management include a requirement to balance supply and demand at the at the sub-basin level for sustainability and disaster risk reduction, while recognizing and protecting downstream users. These targets must permit the tailoring of actions to national realities.

The report argued that agriculture sector must be held more accountable for water use efficiencies and other system efficiencies which limit water demand while maintaining or increasing productivity, ensuring that women and small scale farmholders are provided with the knowledge and technology to be able to play their part, thereby increasing income above poverty thresholds. The energy sector must also be held accountable for water efficiencies in energy and a transition to clean energy, including hydropower, which does not compromise water quality, environmental integrity, community access, or disaster mitigation.

The report was also very clear about the importance of involving the private sector in the global effort to achieve sustainable development goals with respect to water. Water-dependent companies have a key role to play, the report noted, in financing and implementing sound water, sanitation and wastewater management strategies and must step up to the plate or risk significant losses. This is no longer simply corporate social responsibility but sound economic strategy. National governments, multi-national corporations, and international institutions must work together to identify and implement strategies to equitably free up available existing resources. Current expenditures must be more efficient, freeing up and increasing returns on existing resources through integration of inter- and intra-sectoral activities that take advantage of economies of scope and scale. Subject to rigorous due diligence, national governments must identify, explore, and utilize new and emerging financial sources.

Dr. Zafar Adeel, Director of the United Nations University Institute of Water, Environment and Health and Jong Soo Yoon, Head of the UN Office for Sustainable Development, stated that the report filled a critical gap in understanding the complexities associated with water resources and their management, but also provided substantive options that enabled the world to move forward with the global dialogue on the relationship to water and sustainability.

1.4 The 2030 UN Transforming Our World Sustainable Development Agenda

Such reports make it clear that it is not unreasonable to say that water in-security has a very human face. As a global society we face some very substantial and very complex immediate threats to the sustainable presence of the global social order as it exists today. Threats that we have brought on ourselves. But within these challenges resides opportunity. The opportunity before us is humanity's big chance to get it right for future generations. In responding to the urgency and the opportunity of finally getting sustainable development right, the United Nations announced its long anticipated new framework for global action. Launched in New York in the September of 2015, the 2030 *Transforming Our World* agenda (UN 2015b) promises to be the most comprehensive and inclusive effort to positively change the world in all of human history. It was heralded at the time of its release as nothing less than a charter for people and the planet for the twenty-first century. While it remains to be seen if it meets these high expectations, there is no question that the 2030 *Transforming Our World* agenda raises the ceiling on sustainability globally. While it did not receive the same attention in the media, the announcement of 2030 *Transforming Our World* agenda was at least as important as the later climate negotiations in Paris if only because it deals with damage we are doing to other elements of the Earth system that are exacerbating and being exacerbated by climate change. The *Transforming Our World* agenda recognizes that unless we address the problems that form the backdrop to the climate threat, it will not be possible to prevent runaway changes in Earth system function that could bring the conditions that make life possible on this planet as we know it today to an end.

The 2030 *Transforming Our World* agenda is constructed around five themes: people, planet, prosperity, peace and partnership. It is also important to note that this agenda applies equally to the developed world as it does to developing nations. In this context we may wish to remind ourselves of what sustainable development is commonly held to mean. Sustainable development refers to development that meets the needs of the present without compromising the ability of future generations to meet theirs. More specifically, sustainable development recognizes that eradicating poverty in all its forms, combatting inequality within and among countries, preserving the planet, creating sustainable and inclusive economic growth, and fostering social stability are linked to each other and are inter-dependent. Our hope for achieving sustainable development globally resides in the balance between urgency, capacity and will to succeed as demonstrated by each and every UN Member State in making action possible through common but differentiated responsibility at the level of each nation state. It is at the national level that these goals must be met. The degree of our success will depend on governance, which in this context refers to the way in which authority is organized and executed in a society.

The 2030 *Transforming Our World* agenda not only demands far more focused national action within an enhanced framework of global cooperation and coordination, it also underscores the urgency of such action. The agenda makes it clear that we will not be able to deal with the degree of hydro-climatic change we are now witnessing on a global scale unless we are able to translate the 2030 sustainable development agenda to action at the national level. In other words, we won't achieve the goal of sustainable human existence at any meaningful level of prosperity unless we all take common global goals seriously and implement meaningful and measurable actions at the national level in every country in the world, now. This means there can be no laggards particularly in the developed world. It also means that the world cannot afford to leave anyone behind.

So the question becomes one of how any given country goes about integrating sustainable development goals into its national development strategies. This will be particularly difficult in developing countries where the national government has limited means to act or in places incapacitated by conflict. It may be just as difficult, however, in developed countries where national governments have little interest in such strategies or choose to simply devolve responsibility for sustainability without attendant resources to sub-states or provinces which in turn devolve responsibility to cities, towns and rural districts. The 2030 *Transforming Our World* agenda holds that the manner in which we effect the translation from global threat to national action has to be seen as an urgent immediate priority for everyone fortunate enough to be able to be in a position to give meaningful consideration to the future viability of their nation and the communities within their nation in which they live.

One of the ways to re-energize the conversation about sustainable development and humanity's need for resilience in the face of rapid change is to talk about something none of us can live without: water.

Of the nine critical Earth system boundaries which we dare not cross, water plays a significant role in seven. There are 17 goals in the 2030 *Transforming Our World* sustainable development agenda. While important in its own right, water plays a role in 13 of the other 16 goals including ending poverty and hunger; creating healthy lives; gender equity; energy security; promoting economic growth; resilient infrastructure, sustainable cities and economies; and taking action on climate change. Goal 6, however, pertains specifically to water.

The global goal with respect to water is to achieve universal and equitable access to safe and affordable water and sanitation for all. The goal also aims to improve water quality by reducing pollution, halving the proportion of untreated wastewater and substantially increasing recycling and safe reuse globally. The *Transforming Our World* agenda also seeks to protect and restore water-related ecosystems in part by implementing integrated water resource management at all levels including transboundary basins by 2030.

The world learned from the earlier Millennium Development Goals that we need to better address the multiple roles water plays in establishing, maintaining and improving the human condition. We need to take advantage of the synergies that exist between effective water management and benefits that accrue not only directly to human health and well-being but also to the environment and the economy.

We are very good at evaluating the state of water resource use, determining ecosystem health and evaluating potential climate impacts but we are less capable of changing our practices once we have characterized those parameters. Creating a systems approach to managing water has to be seen as synonymous with sustainability and resilience.

The 2030 *Transforming Our World* agenda focuses considerable attention on the not insignificant matters of cities. Some 92% of the population growth which has brought the last 1.2 billion into the world has occurred in cities. Some 60% of the urban space required to accommodate future populations has yet to be built. Sustainable development goal 11 aims to make the world's cities and human settlements inclusive, resilient and sustainable.

If we are to live sustainably on this planet, cities must commit to achieving the goals of the 2030 *Transforming Our World* agenda. This means that resilience has to be seen as a child of sustainable development. While the *Transforming Our World* agenda does not provide enough guidance to help any given city in terms of specific pathways to resilience, it does provide clear recognition of the critical importance of cities with respect to the goal of creating a sustainable civilization. While cities can certainly act in any manner consistent with their own local needs and vision with respect to resilience what they do must contribute to the global sustainability effort. Sustainable development in cities means adequate, safe and affordable housing and basic services for all as well as safe, affordable transportation systems. It also means strengthening and safeguarding cultural diversity.

The *Transforming Our World* target of providing universal access to safe, inclusive and accessible green and public spaces is consistent with all efforts to improve resilience especially in cities vulnerable to heat island effects. Rising levels of urban heat are now seen to constitute the single greatest climate-related threat to human health globally. Urban heat waves now account for more deaths per year globally than all other forms of extreme weather events. We don't have to wait for the future to witness this. It's happening now. The *Transforming Our World* vision is that by 2030 sustainable, resilient cities will significantly reduce the number of deaths and the economic and psychological effects caused by disasters including water-related catastrophes. The target is that as early as 2020 we will have substantially increased the number of cities and human settlements adopting and implementing plans to mitigate and adapt to climate change and to enhance resilience to disasters. The template for holistic disaster risk management at all levels is the United Nations supported Sendai Framework for Disaster Risk Reduction.

By 2030 we also want to reduce the adverse per capita impact of cities on surrounding regions. This target of course speaks directly to air quality, water contamination and other waste management issues. But the agenda does not stop there. Another global sustainable development target that has great relevance here is recognition of the need to support positive economic, social and environmental links between urban, peri-urban and rural areas through the strengthening of national and regional planning.

The 2030 *Transforming Our World* global sustainable development agenda makes it very clear however that sustainable development can no longer simply aim for environmentally neutral solutions. If we are to achieve any meaningful level of sustainable development all development has to not only be sustainable, but restorative. Nowhere is this truer than in agriculture.

As has been noted by many UN agencies and underscored at the United Nations Conference on Trade and Development in 2013, over the past 50 years farming has come to be controlled by the demands of machinery rather than the nature of agricultural lands. Universities, corporations and governments continue to almost unanimously support industrial agriculture despite the now almost overwhelming evidence of its impacts. These impacts include soil erosion and salinization, aquifer depletion, and dependence of fossil fuels and toxic chemicals. The damage caused by industrial agricultural practices also includes its polluting effect on streams, rivers and lakes; the loss of genetic and biological diversity it causes and the destruction of rural communities and cultures of animal husbandry that follow in its wake. We now realize that, in addition to these problems, climate change has the potential to irreversibly damage the fundamental resource base upon which agriculture depends, with grave consequences for food security globally.

What we need is another agricultural revolution. We need to focus on critical inter-dependencies especially as they relate to water, food and energy security. Reliable, secure access to water is a necessary condition for food security. One of the *Transforming Our World* goals pertaining to ending hunger that relates directly to the food-growing regions of the world is to implement agricultural practices that increase productivity and production while at the same time help maintain ecosystems, strengthen capacity for adaptation to climate change and reduce the impacts of extreme weather, drought, flooding, and other disasters that progressively diminish land and soil health. The goal is to do this by 2030.

By 2030, we also have to halve per capita global food waste at the retail and consumer level and reduce food losses along production and supply chains, including post-harvest losses.

By 2030, the world also needs to achieve sustainability with respect to management and efficient use of natural resources. We have to figure out how to manage chemicals and wastes throughout their life cycles. We have to substantially reduce waste generation. People everywhere have to have the relevant information and awareness of what they can do in terms of their own lifestyles to help themselves and their communities achieve sustainability. We have to make the transition from seeing waste as waste to seeing waste as wealth. If we don't want these kinds of problems to get away on us we have 15 years to do this.

Regrettably, the fact that is emerging is that nothing we do will be sustainable unless we address the climate threat. If we are to have any hope of achieving a meaningful level of sustainability globally, we have no choice but to combat climate disruption and its impacts particularly as they relate to water. To achieve that end every country in the world has to integrate climate change into its national policies, strategies and planning processes. We could start down that road by rationalizing inefficient fossil-fuel subsidies that encourage wasteful consumption by removing market distortions.

It is important to be very clear on this point: climate impacts will affect the development trajectory of all nations, rich and poor. In fact, there is a proven link between climate disruption and de-development. While the link between recurring extreme weather events and challenges of maintaining critical physical and social infrastructure in developing countries has been noted by organizations like the United Nations and the World Bank, the extent to which climate disruption has begun not to just slow, but reverse economic development was not recognized as a global economic threat until 2016. In January of that year, the World Economic Forum held in Davos, Switzerland released its annual Global Risks Report. As part of that survey nearly 750 experts assessed 29 separate global risks for both impact and likelihood over a 10 year time horizon. The risk with the greatest potential impact in 2016 was found to be the failure of climate change mitigation and adaptation. It is important to note that this was the first time since the report has been published that an environmental risk had the top ranking. The failure to mitigate and adapt to climate change was now considered to have created greater potential risk of damage than weapons of mass destruction (2nd); a global water crisis (3rd); large scale involuntary migration (4th) and the economic consequences of energy price shock (5th).

When the risks were ordered in terms of likelihood alone, the number one risk in 2016 was thought to be large-scale involuntary migration, followed by extreme weather events (2nd); failure of climate change mitigation and adaptation (3rd); interstate conflict with regional consequences (4th) and major natural catastrophes (5th). It was also noted that such a broad risk landscape was unprecedented in the history of the publication of the report. The authors also noted that this more diverse risk landscape came at a time when the toll from global risks appeared to be on the rise.

The report went on to note that the global climate warmed enough in 2015 to raise the global average surface temperature to the milestone of 1 °C above the pre-industrial era for first time in human history. The report also observed that according to the UN's Refugee Agency, the number of people around the world forcibly displaced in 2014 stood at 59.5 million, which is almost 50% higher than the number of refugees on the move at the beginning of World War II in 1940. The report also noted that available data supported the likelihood of increasing risk annually in all

24 of the categories which the World Economic Forum had been measuring since 2014.

In addition to measuring their likelihood and potential impact, the World Economic Forum Global Risks Report 2016 also explored interconnections among risks. The report found that a very small number of risks accounted for interconnections than in 2015. The two risks that were most interconnected in 2016 were profound social instability and structural unemployment or under-employment which accounted for 5% of all interconnections. The report quoted Margareta Drzeniek-Hanouz, Head of the Global Competitive Risks program at the World Economic Forum, who underscored the importance to leaders of understanding such connections as a means of prioritizing areas for action as well as planning for contingencies. "We know climate change is exacerbating other risks such as migration and security, but these are by no means the only interconnections that are rapidly evolving to impact societies, often in unpredictable ways," she said. "Mitigation measures against such risks are important, but adaptation is vital," she concluded.

The report noted that in terms of risk, it is becoming difficult to determine which panic button to press. Cecilia Reyes, the Chief Risk Officer of the massive Zurich Insurance Group, probably explained the circumstances as they stood in 2016 best at least in terms of impacts on global economic performance. "Climate change," she said, "is exacerbating more risks than ever in terms of water crises, food shortages, constrained economic growth, weaker societal cohesion and increased security risks. Meanwhile geopolitical instability is exposing businesses to cancelled projects, revoked licenses, interrupted production, damaged assets and restricted movement of funds across borders. These political conflicts are in turn making the challenge of climate change all the more insurmountable – reducing potential for political co-operation as well as diverting resources, innovation and time away from climate change resilience and prevention."

A haunting image of the extent and nature of contemporary risks was put forward at the 2016 World Economic Forum by the global insurance giant Munich Re. The image was a map of interconnections between various economic, environmental, geopolitical, societal, and technologic risks associated with the failure to effectively and meaningfully adapt to climate change. What the map illustrates was the cascading effect of the failure to adapt to hydro-climatic change. On a global scale failure leads first to greater vulnerability to extreme weather events; food crises; water crises; large-scale involuntary migration; further man-made environmental catastrophes which in turn lead to biodiversity loss and Earth system collapse.

This isn't speculation; it is already happening. At the Davos Forum it was noted that the terrible violence rocking Syria and the spill-over effects in Europe in 2015 did not start as a war, it started as a 5-year drought that contributed to sparking a war. What we haven't understood until now is the extent to which the fundamental stability of our political structures and global economy are predicated on relative hydrologic predictability. As a result of the loss of relative hydrologic stability, political stability, and the stability of our global economy in a number of regions in the world are now at risk. We are only now beginning to understand how complex this issue has become.

Even though it often sucks all the air out of the room, climate disruption is only one of the 19 global sustainability challenges that we need to address through the 2030 *Transforming Our World* sustainable development agenda. We have to better protect our oceans first from land-based activities including marine debris and nutrient pollution, protect and restore coastal ecosystems, regulate harvest, and end overfishing; and halt perverse subsidies and destructive fishing practices. There is urgency in this. If we don't minimize and reverse the impacts of ocean acidification by 2030, we will lose one of humanity's most important sources of food and livelihood.

To achieve these goals we must increase scientific knowledge, further develop research capacity and stimulate faster transfer of marine knowledge and technology.

The UN has also set 2030 as the goal for combatting desertification, restoring degraded land and soil, halting the degradation of natural habitats; minimizing the impacts of invasive species, and halting the loss of natural biodiversity with the aim of ensuring that sustainable development remains possible in the future.

We cannot achieve these goals without making the world a safer place. In order that sustainable development is not slowed or reversed we have to significantly reduce all forms of violence and related deaths everywhere. We must, all of us, promote the rule of law at all levels; and reduce corruption and bribery in all their forms.

Our sustainability also depends of reducing illicit financial and arms flows, and combatting all forms of organized crime. We must demand responsible, inclusive, transparent, participatory and representative decision making wherever we live.

Looking at this long list of goals it becomes very clear that fulfilling the *Transforming Our World* agenda will require that we find new ways to pay for what we need to do. We will need to re-prioritize domestic and public funding; domestic and international business finance and international development cooperation to find that kind of money. Subject to rigorous due diligence, national governments must identify, explore, and utilise new and emerging financial sources. It will be critical to encourage and promote effective, public, public-private and civil society partnerships in efforts to develop new kinds of financing and resourcing strategies.

All 17 of the sustainable development goals and all 169 of the targets in the 2030 *Transforming Our World* agenda are held to be universal, indivisible, and interlinked and as such should all be regarded as having equal importance and be accorded equal priority with respect to implementation. While the agenda respects each country's policy space, it is recognized that national development efforts need to be supported by an enabling international economic environment. That environment must include coherent and mutually supportive world trade, monetary and financial systems and strengthened and enhanced global economic governance.

The very real specter of unanticipated new issues emerging between now and 2030 is virtually a given. We will need to expand the data available upon which to make sustainable development decisions to include broader satellite-based Earth system observations and geospatial information.

Implementation of the *Transforming Our World* agenda will be tracked to make sure no one is left behind. The goals and targets will be followed and reviewed using a set of global indicators which are scheduled to be put in place by March of 2016.

All reviews of progress toward implementation of the agenda must be country-led and country-driven. They should be regular and inclusive, draw on observations and contributions by Indigenous peoples, civil society, the private sector, and other stakeholders. Ideally such reviews should provide a new platform for partnerships.

There is a lot to do. Is there economic opportunity in pursuing these goals and targets? Absolutely. If we want to have meaningful and prosperous lives in the coming decades, achieving these goals now must become a pillar of every economy nationally and globally.

Theoretically, all the elements required to create sustainability are included in the agenda. The great challenge and urgency is to make these goals and targets a priority at the national level.

To this end, all member states are encouraged to develop ambitious national responses to the implementation of the *Transforming Our World* agenda as soon as it is practicable. This does not by any means suggest starting over; what this means is building on and focusing existing planning instruments, sustainable development, and resilience enhancement strategies.

1.5 Why Careful Water Management & A Stable Climate Are Critical to Sustainability

The more we focus on climate change the more the focus is on water, in particular water security. The era in which we live might be called the storm after the calm. After a period of relative hydro-climatic stability during which we created most of our built environment, step-like changes to our hydro-climatic circumstances are demanding that we redefine what development and sustainability mean not just nationally, but globally. This in turn demands that we reassess personal and collective vulnerability, accountability, and liability and adapt quickly to changing circumstances if we want to sustain our prosperity in the face altered hydro-climatic conditions.

While they all must be linked together, some sustainable development goals must be achieved before it will become possible to achieve others. Climate action is one goal that if not achieved, will make it difficult if not impossible to achieve many of the others. The composition of the atmosphere is the lynch-pin holding the ice-water-weather-climate system of the planet together. If we cannot stabilize the composition of the Earth's atmosphere, we cannot know to what end our sustainability efforts must aim because we will not know the conditions to which we will need to adapt. Because of the warming effect on the global hydrologic cycle, we have to stabilize the composition of the atmosphere or we will not be able to achieve the goal of providing clean water and sanitation for all.

Without stabilizing the composition of the Earth's atmosphere, we will no longer be able to reliably predict where our food will come from. What happened in California recently and in Syria between 2006 and 2011 is beginning to happen more widely and will continue to do so for the foreseeable future, threatening food security everywhere. But we don't just grow food on land; without stabilizing the composition of the Earth's atmosphere we cannot reverse the acidification that is threatening ocean food webs. What we learn from this is that we cannot have food security without climate security, and without food and climate security there cannot be peace and justice for all.

But there is more; our cities were designed for climatic circumstances that will soon no longer prevail. Without stabilizing the composition of our atmosphere, we will not know to what standards we need to redesign our cities in the face of ever more powerful storms, bigger floods, and longer heat waves. The complicated challenge of accurate prediction aside, we cannot know how much sea level rise we need to protect our cities against. Without stabilizing the composition of the atmosphere, urban resilience will remain a moving target constantly beyond our grasp. Nor can we, without stabilizing the composition of the Earth's atmosphere, prevent desertification, halt ecosystem collapse, or slow the accelerating rate of extinction of the other creatures with which we share this planet. It is now very clear that the relative climatic stability we have enjoyed over the past century or so will not return for centuries, if ever. Failure to realize this could cost us our prosperity. It could even cost some of us our lives.

It is clear also we need to define a safe place in terms of sustainability to which all of humanity must aim. In this quest knowledge is not enough; we need the will and the permission to solve global change problems. That is why COP 21 in Paris was so important. We need the will and the permission because, as the motivating theme of COP 21 made clear, "Later, it will be too late." So how did we do in Paris?

1.6 Separating the Hype from the Hope in Paris: The Hype

Perhaps it was British journalist George Monbiot who said it best. What Monbiot essentially said was that by comparison to what it could have been, what happened in Paris was a miracle. Compared to what it should have been, however, it was a disaster. As American climate change action advocate Bill McKibben pointed out we are no longer sitting around the table negotiating with other countries. We are dealing with fundamental atmospheric physics, and the physics holds all the best cards. It means that if temperatures on Earth rise 4–5 °C, significant areas of the planet are likely to become uninhabitable for at least parts of the year.

The Paris Agreement (UN 2015a) continues to represent a linear approach to the problem, leaving all of humanity vulnerable to unanticipated step-like changes in our climate-related circumstances. One does not have to read far into the agreement to realize we have wasted a generation in responding to the climate threat. There are many things in the Paris Agreement that we should have done already such as meaningfully reducing greenhouse gas emissions, developing methodologies for assessing adaptation needs with a view to assisting developing countries; strengthening regional cooperation on adaptation; climate-proofing national and regional economies; and developing integrated approaches to averting, minimizing and addressing large-scale displacement of people as a consequence of climate disruption.

The Paris Agreement allows for examination of the risks of damage and displacement but it does not allow attribution of blame, suggestions of liability, or recommendations of compensation. If you are an island state in the process of being submerged there is no one you can hold directly accountable.

Because it has taken us 20 years just to agree upon goals, pace has now been identified as everything. That said, the Paris Agreement grants the world five more years not to set the pace as might be expected but just to set national emissions reductions targets. Prior to 2020, nothing is obligatory. The degree to which this agreement is largely aspirational is obvious in that it calls for the establishment of a mechanism for tracking emissions at a national level but maintains that even these contributions to global reduction targets are still voluntary. The agreement "recognizes" the important role of providing incentives for emissions reduction activities including tools such as domestic policies and carbon pricing; but that's all.

The agreement rightly suggests that governments cannot and should not be allowed address the climate threat by themselves and "welcomes" the involvement of the private sector, civil society, financial institutions, cities, and other sub-national to join the initiative. That the agreement did not say that it "expects" the participation of these entities or even "demands" participation of these entities in implementing solutions underscores its aspirational character.

Then there is the not insignificant issue of timeframes. The first global stocktaking of global implementation progress will not take place until 2023. Even participation in mechanisms for establishing targets for greenhouse gas emissions reductions beyond 2023 into mid-century and longer is voluntary. In a very real sense, these conditions in themselves undermine the agreement from its very inception. In the 5 years Parties to the Convention have granted themselves to establish how much they will contribute to global emissions reductions we could very well blast past any real opportunity to limit mean warming to 1.5 °C. The delay in action could even put the 2 °C target out of reach. All self-congratulations aside it appears that once again all we have done is kick the can down the road. Cut the numbers any way you want but at the end of the Paris conference we still have less than a 50% chance of avoiding runaway climate impacts.

The agreement also has other shortcomings; there is no mention of direct human health risks. Aviation and shipping are not mentioned or included in the Paris Agreement. Water security is not mentioned in the agreement even though effective management of water is the foundation of climate security.

The agreement focuses largely on technology and technological transfer while largely ignoring the improvement of ecosystem function in service of keeping the world from warming more than 1.5 °C.

While the agreement recognizes the critical need to reduce emissions generated by deforestation and forest degradation, there is no reference beyond forestry management to ecosystem-based mitigation and adaptation strategies such as the enhancing of soil health as a means of increasing carbon sequestration. In this agreement, agriculture gets a "get out of jail free" card. Nor is it clear where the money is going to come from to finance even what has been proposed in terms of climate action. Though much is made in the agreement about transparency, it allows developing countries to weasel out of transparency reporting by making such reporting at the level of in-country reviews optional.

The agreement relies on transparent, non-adversarial, non-punitive compliance. It is also an agreement that is easy to get out of. At any time after 3 years from the date on which the agreement comes into force for a signatory, that signatory can withdraw from the process by simply giving written notification of the intent to do so. The withdrawal will take effect 1 year after the filing of notification.

There is also some question as to whether this agreement can withstand outside disruptions such as large-scale terrorism events and cyclic economic collapses neither economists nor politicians appear to be able to predict or control. Because it is non-binding, the agreement is also highly vulnerable to political manipulation. The question then becomes whether it can survive political turmoil or the election of radically different new governments. How, for example, will the agreement fare now that Donald Trump has been elected President of the United States?

The biggest failing of the Paris Agreement, however, may reside in the fact that it offers only a long list of urgings, invitations, and encouragements to signatories to act now and in the future. But there is nothing to force the signatories to do so. At this time the agreement merely "urges" parties to make voluntary contributions to the timely implementation of the process to which they agreed. We are still working on reaching the peak of greenhouse gas emissions with the promise that once we have done so we will start cutting them. In this knowledge, it is hard to be hopeful. But there is hope.

1.7 Separating the Hype from the Hope in Paris: The Hope

The real miracle in Paris is that for a moment at least we got 193 nations – rich and poor – to agree something, however limited. That is an achievement in itself. Whether everyone will continue to agree once the delegations return home and are worked over by the divided political constituencies remains to be seen.

No, it's not perfect – in fact it is far from perfect – but now at least we have something to build on – something concrete we can work towards together that didn't exist until COP 21. This is a beginning, not an end.

The first thing that is really important about the Paris Agreement is that it nests climate action within the larger context of the UN's 2030 *Transforming Our World* global sustainable development agenda as well as other important UN conventions.

It has been said often that addressing the global climate threat will require nothing less than the kind of vision, concentration of finances, resources, intelligence, and purpose that permitted the United States to put a man on the moon in 1969. In nesting the climate challenge within the expanded global dialogue concerning the sustainability of human presence on this planet, this agreement recognizes that what we have arrived at is the need for multiple moon-shots in each of the 17 areas critical to sustainability and that we have to urgently embark upon all simultaneously.

Within this context, the agreement clearly recognizes that climate change represents an urgent and potentially irreversible threat to human societies and the planet and thus requires the widest possible cooperation by all countries, and their participation in an effective and appropriate international response, with a view to accelerating the reduction of greenhouse gas emissions. It also recognizes that deep reductions in global emissions will be required in order to achieve the ultimate objectives of avoiding dangerous anthropogenic warming.

The agreement importantly recognizes the need for an effective and progressive response to the urgent threat of climate change on the basis of the best available scientific knowledge. The agreement makes it very clear that the global effort must be to hold the increase of global average temperature to well below 2 °C and to limit the increase to 1.5 °C, if possible. In establishing the 1.5 °C target, the agreement recognizes the specific needs and special circumstances of developing countries and countries particularly vulnerable to harmful climate effects. The agreement also conversely recognizes that many nations may be affected not only by climate change, but also by the impacts of measures taken in response to it. The agreement also clearly recognizes the fundamental priority of safeguarding food security and ending hunger, and the particular vulnerabilities of food production systems to the adverse effects of climate change.

While acknowledging that climate change is a common concern to all of humankind, the agreement also makes it clear that actions taken to minimize its effects must be respectful of human rights, the right to health and the rights of indigenous peoples as well as gender equality and intergenerational equity.

Within these parameters, the agreement calls for the setting and achieving of economy-wide absolute emission reduction targets at the national level. Parties to the agreement have to formally submit their targets for greenhouse gas emissions reductions no later than 2020 and resubmit revised targets every 5 years thereafter. Signatories to the agreement are bound in such submissions to clearly and transparently include common baseline references such as the year to which emissions reductions must be compared; methods utilized in estimating and accounting for anthropogenic emissions; assumptions and methodological approaches in arriving at targets; timelines for implementation; and explanation of how their reductions of emissions contribute to the objective of strengthening the overall global response to the threats of water-in-security and climate instability. The common methodologies for accounting will be established by the Intergovernmental Panel on Climate Change and will require that parties do not double count or arbitrarily fail to include carbon sinks or sources.

Article 5 of the agreement recognizes the importance of preserving and enhancing carbon sinks and provides clear marching orders in this regard for forestry management. The agreement encourages signatories to take action to implement and support policy approaches and positive incentives for activities relating to reducing emissions from deforestation and forest degradation, and enhancing the role of conservation and enhancement of forest carbon storage in support of sustainable forest management.

The agreement also makes it clear that adaptation is urgently required. The agreement calls upon signatories of the agreement to ensure that education, training, increased public awareness, participation, and improved access to information are adequately considered at the national level in capacity-building associated with strengthening the global response to the climate threat. As noted above, the agreement also acknowledges that governments can't do all of this alone. The achievement of any meaningful level of water and water-related climate security will require the coordinated action of all official Parties as well as non-party stakeholders including civil society, the private sector, financial institutions, cities and other subnational jurisdictions, local communities, and indigenous peoples.

Though still non-binding, Article 7 on Adaption, which calls for formalizing and implementing national adaptation strategies, is very strong. This article makes it clear that adaptation action should follow a country-driven, gender-responsive, participatory, and fully transparent approach, taking into consideration vulnerable groups, communities and ecosystems, and should be based on the best available science and, as appropriate, traditional knowledge. It also calls for sharing of information, good practices and lessons learned and the further strengthening of scientific knowledge on climate, including research, systematic observation of the climate system and the development of early warning systems that will inform and support decision-making.

Article 8 establishes the role of the UN's 2030 *Transforming Our World* sustainable development goals in averting, minimizing, and addressing loss and damage from extreme weather events.

What happened in Paris should be of particular interest to the private sector. The agreement calls for the enhancement of linkages and creating synergies between mitigation, finance, technology transfer, and coordination of non-market approaches to sustainable development. This agreement is all about opportunity linked to hastening the transition to renewable energy.

The agreement also makes it very clear that climate security cannot be achieved without the cooperative engagement of the average citizen in tandem with the full support of the private sector. What was implied but not said in the agreement was that individuals and corporations with amassed wealth are going to have to put that wealth to work in service of the planetary good if they are to protect the sources of that wealth or to ensure that wealth has the same meaning in a massively changed world.

But even the most optimistic interpretation of the Paris Agreement has to be tempered by a realistic judgment of human nature. Though we are doing an ever better job of characterizing and depicting the damage we are doing to the biodiversity-based planetary life support system upon which we depend to make our civilization possible, we appear incapable, at the moment at least, of adequately slowing that damage. After decades of tough sledding with respect to advancing climate change mitigation and adaptation, we find ourselves starting all over again from only 100 m further down the road to meaningful action. We can only hope that this time, the results will be different. The first test of the durability of the Paris Agreement will be to see if Parties to the agreement actually ramp up their carbon reduction programs between now and 2020. The second test will be to see if the 100 billion dollar climate fund is topped up by 2020. The final test is whether we will be able to keep our Paris promises and build on them beyond 2020.

"The climate conference in Paris," the French Ambassador to Canada Nicolas Chapuis said in Ottawa 3 weeks before the conference began, "is an opportunity to put out the fire that is burning our house down." Flames were no longer seen to be shooting out of the roof at the close of the conference, but clearly the fire is still smoldering inside our house. Hope remains, however, that this fire can be extinguished.

In Mid-February of 2015, the World Bank launched a report about the cost of meeting the *2030 Transforming Our World* Sustainable Development Goals 6.1 and 6.2 (Hutton and Varuguese 2016) which respectively address targets linked to water and sanitation. The three major findings of the report were encouraging. The first major finding was that current levels of financing can cover the capital costs of achieving universal basic service for drinking water, sanitation, and hygiene by 2030, provided resources are targeted to the needs. The second major finding was capital investments required to achieve the water supply, sanitation, and hygiene related to targets 6.1 and 6.2 amount to about three times the current investment levels. Finally, the report observed that sustained universal coverage requires more than capital inflows: financial and institutional strengthening will be needed to ensure that capital investments translate into effective service delivery.

The Paris Agreement is infinitely better than what we were left with after Copenhagen but despite fine words and high aspirations, when you separate the real hope from the hype what is missing is a binding common commitment to act. We are on the river of no return and urgently need clear action now to invest the Paris Agreement with the substance it presently lacks. As with all of the UN's *2030 Transforming Our World* sustainable development goals, that substance has to manifest itself at the national and sub-national level. It is at the national and subnational level globally that all hope of addressing threats to global water security and climate stability presently reside. It is what happens now at the national and subnational level that will put a human face on water security. There is a lot at stake. The failure to act now – while there is still an opportunity to do – will mean that the bodies of more children like Alan Kurdi will wash up on lonely beaches in a world in which water insecurity will lead to greater regional tensions, conflict and involuntary human migration.

References

Hutton, G., & Varughese, M. (2016). *The costs of meeting the 2030 sustainable development goal targets on drinking water, sanitation and hygeine.* World Bank Group. Available at http://www.wds.worldbank.org/external/default/WDSContentServer/WDSP/IB/2016/02/08/090224b08415bdcc/1_0/Rendered/PDF/The0costs0of0m0ene000summary0report.pdf

InterAction Council. (2012). *InterAction Council celebrates 30 years with a gathering in Tianjin, China.* Available at http://www.interactioncouncil.org/interaction-council-celebrates-30-years-gathering-tianjin-china

Schuster-Wallace, C. J., & Sandford, R. W. (2015). *Water in the world we want.* United Nations University Institute for Water, Environment and Health and United Nations Office for Sustainable Development. Available here http://inweh.unu.edu/wp-content/uploads/2015/02/Water-in-the-World-We-Want.pdf

UN. (2015a). *Adoption of the Paris agreement, FCCC/CP/2015/L.9*. Available at https://unfccc.int/resource/docs/2015/cop21/eng/l09.pdf

UN. (2015b). *The United Nations 2030 transforming our world global sustainable development agenda, A/RES/70/1*. Available at https://sustainabledevelopment.un.org/post2015/transformingourworld

UNCTAD. (2013). *Trade and environment review 2013 – Wake up before it is too late: Make agriculture truly sustainable now for food security in a changing climate*. Available at http://unctad.org/en/pages/PublicationWebflyer.aspx?publicationid=666

Chapter 2
Water Security as the Centerpiece of the Sustainable Development Agenda

Zafar Adeel

Abstract The 2030 Agenda for Sustainable Development has adopted some unique fundamentals: it is universal and applies to all countries; it is comprehensive and seeks to completely eliminate problems; it is complex as shown by the large numbers of goals and targets; and, it is ambitious as it aims to fix major global problems in a 15-year span. It signifies a paradigm shift in international development that emphasizes the role of national governments and other domestic stakeholders. Water security is a keystone element in achieving the 2030 Agenda – not just for Sustainable Development Goal (SDG) 6 that focuses primarily on water security, but a number of targets embedded within other SDGs related to health, cities, consumption, marine resources, and terrestrial ecosystems. The concept of water security is quite pervasive in the international discourse on sustainable development, and lines up strongly with the notion of water as a human right. For the SDGs to succeed, water security must be achieved across all geographic scales – from international to national to sub-national to community – and across all social strata. Achieving this objective requires considerable alignment of all stakeholders: governments, farmers, private businesses, civil society organizations, researchers, and scientists. Implementation to achieve the various water-related targets offers an excellent opportunity for creating innovative and integrated solutions. These solutions require equally innovative and enabling policy environment, a role that is typically well suited to governments at various levels. Future outlook for achieving universal water security for all individuals is positive, but requires considerable rethinking around economic and social development.

Keywords Sustainable development • SDGs • Water security • Human development • Enabling policies

Z. Adeel (✉)
Pacific Water Research Centre, Simon Fraser University, Burnaby, BC, Canada
e-mail: zadeel@sfu.ca

D. Devlaeminck et al. (eds.), *The Human Face of Water Security*, Water Security in a New World, DOI 10.1007/978-3-319-50161-1_2

2.1 Introduction

A general correlation between development and security stems from the arguments put forward by the United Nations Development Programme (UNDP) in which it states that human security can only be achieved through human development that is provided by viable and economically stable states (UNDP 1994); this notion of security – also referred to as "human well-being" in other contexts – includes basic materials for good life (such as secure and adequate livelihoods, sufficient food, shelter, clothing), good health (such as a healthy environment, clean air and water), good social relations, personal security, security from manmade and natural disasters, and freedom of choice and action (MA 2005). It is well recognized that the international community – loosely defined as the collection of UN organizations, bilateral aid agencies, international development banks, international non-governmental and civil-society organizations, donor organizations, and foundations – has to play a major role in achievement of human security. The evolving role of the international community in achieving development objectives led, in the late 1990s, to the development of "Poverty Reduction Strategy Papers" for most developing countries (World Bank and IMF 2005). Similarly, the Millennium Declaration by the world leaders in September 2000, followed by the adoption of the Millennium Development Goals (MDGs), provided a global framework in which sustainable development could take place and human security could be achieved (please see Annex 2.1 for details of MDGs).

This chapter provides evidence to support the argument that water is a building block for human security; that is, human security is closely tied to water security. It is worth noting that water security is a broad concept that encompasses the more traditional notions of security pertaining to armed conflict, civil wars, and outright wars but at the same time expands to include a more anthropocentric definition: it is a basic and fundamental element of human wellbeing (Adeel 2012). This latter approach is commonly used throughout this volume and fits well with the declaration of water and sanitation as a fundamental human right by the United Nations General Assembly in July 2010 (UNGA Resolution A/Res/64/292). A jointly-developed working definition by the United Nations describes the ingredients of water security as: "… *the capacity of a population to safeguard sustainable access to adequate quantities of acceptable quality water for sustaining livelihoods, human well-being, and socio-economic development, for ensuring protection against water-borne pollution and water-related disasters, and for preserving ecosystems in a climate of peace and political stability*" (UN-Water 2013). To state the obvious, the elements contained within this definition of water security overlap closely with the ones described earlier as ingredients of human security and wellbeing. One can, therefore, explore how achieving water security can contribute to overall human development, whether this development is defined in economic, cultural, or social domains.

When discussing the nexus of water and human development, it is important to keep in mind the dual nature of water. It is on the one hand a resource that underpins numerous social and societal activities, and needs to be maintained at a sufficient quality and quantity through effective environmental management and conservation. On the other hand, water is also viewed by many as a "sector" in which protection of human rights requires provisioning of safe water through investments in infrastructure, institutional, and management regimes. While both dimensions of water require full attention to achieve water security, the balance between the two is difficult to achieve in practice – as typically observed in most developing countries. Conversely, it can be argued that all developed countries have invested heavily in water-related institutions and infrastructure, which has resulted in reliable water services, minimized impacts on human health, and reduced risk of adverse impacts from water-related disasters (UNU and UNOSD 2013).

This chapter concerns itself with exploring the nexus between water security and human development. In the second section of the chapter, it does so by investigating the new international development agenda and by exploring how water security fits into the far-reaching declaration and agreements that underpin this agenda. The third section of the chapter tests the argument that water security underpins nearly all dimensions of development and the absence of water security means that not only further development would be stunted but gains made during the last few decades towards achieving human security would also be under threat. The discussion in the fourth section of this chapter unpacks why, despite ostensibly solid economic and evidence-based arguments, achievement of water security has largely been elusive during the MDG era (2000–2015). This narrative sets up the stage for a short discussion on how the 'ship can be turned around' in the fifth section. The role national governments as well as the international community need to play in achievement of water security is discussed and some options for innovative resource mobilization are presented. Overall, it is argued that national governments are the most important actors and facilitators in achievement of water security and they can do so while exploiting the constructs of the development paradigm that has emerged during 2015.

2.2 The Global Context

2015 has turned out to be a year in which the global development agenda was re-cast in a major way. It can truly be termed as a 'blockbuster' year.

Conclusion of the Millennium Development Goals (MDGs) in 2015 was a landmark in itself; many previous publications have offered in-depth analysis and assessment of achievements during the 15-year period (2000–2015) (Fehling et al. 2013; UNU and UNOSD 2013; Fukuda-Parr et al. 2014). Many have argued that a number of accomplishments could not have been achieved were it not for the push towards accomplishing MDGs (UNSGAB 2015). Conversely, some authors have criticized the MDGs for being selective and incomplete, resulting in far less than satisfactory outcomes (UNU and UNOSD 2013). Perhaps Fukuda-Parr et al. (2014) offer the most critical review, stating that the target-setting approach used in MDGs turned out to be a limited and blunt tool for development that diverted attention from other important development objectives, and that poorly selected MDG indicators contributed to a distorted perspective of what development was taking place. Regardless of the success or otherwise of MDGs, the process of their implementation left us with many important lessons. For example, it was concluded that the sustainable development agenda must be characterized and viewed as a universal compact, the formation of which must involve all stakeholders and be based on growth models that decrease inequality and risk (IRF2015 2013). Analysis of the MDG implementation undertaken by the United Nations University Institute for Water, Environment and Health (UNU-INWEH) similarly highlighted that a future framework for sustainable development must address in an integrated manner the issues of poverty eradication, achievement of social and economic equity, environmental and resource sustainability, and opportunities for job creation and economic growth (UNU and UNOSD 2013). The study also pointed to the need for formulating new development goals that are universal in nature, identifying targets that are well-defined and measurable, ensuring that the development goals are sensitive to changes in global drivers, and supporting national and sub-national processes that lead to effective governance mechanisms. It should be noted that impacts of a variety of global drivers need to be considered when setting targets, including but not limited to system-wide changes to climate and hydrology, global economic crises, population growth, escalation in food prices in the international market, and armed and civil conflict.

An extended process of negotiations and dialogue preceded the formulation of a global framework that was eventually termed as the 2030 Agenda for Sustainable Development (UN 2015). Earnestly kicked off in the aftermath of the Rio+20 Summit,[1] the consultation process involved a broad range of stakeholders: national governments, United Nations organizations and agencies, non-governmental and civil-society organizations, scientific and research community, think-tanks, international development organizations and banks, and the general public. According to estimates by the United Nations, over a million opinions were received throughout this consultative process, which is truly unprecedented. Global leaders gathered in New York in September 2015 and signed off on the 17 Sustainable Development Goals (SDGs) and 169 underlying targets. These included SDG 6, which focuses primarily on universal availability of water and sanitation, and includes eight targets that address issues such as access to water and sanitation, water quality, water use efficiency, ecosystems conservation, trans-boundary cooperation, and integrated water resources management (please see Box 2.1 for details of SDG 6). It is noteworthy that the term "water security" does not appear in the language used by the UN in formulating SDG 6, and yet all the constituents of water security are explicitly addressed within the targets of SDG 6. In off-the-record conversations with the author, a number of representatives of the UN Member States have noted that the term "security" was not viewed favorably by their respective governments as the over-arching definition of SDG 6. This aversion may, in part, be due to the fact that

[1] United Nations Conference on Sustainable Development, Rio de Janeiro, Brazil, 20–22 June 2012.

the word "security" does not translate into French and Spanish languages in quite the same way as it is used in English. Nonetheless, it should be noted that for the sake of this chapter, the achievement of water-related SDG targets is considered synonymous with having achieved water security.

Further, a number of water-related targets are embedded within the other SDGs: SDG 3 on health (target 3.3 – combat water-borne diseases; target 3.9 – reduce deaths/illness from water pollution and contamination), SDG 11 on cities (target 11.5 – reduce impacts of water-related disasters), SDG 12 on consumption-production (target 12.4 – sound management of wastes to reduce releases to water), SDG 14 on marine resources (target 14.1 – reduce marine pollution from land-based activities), and SDG 15 on terrestrial ecosystems (target 15.1 – sustainable use of freshwater ecosystems and their services; target 15.8 – reduce impact of invasive alien species on water ecosystems). Overall, it is clear that water serves as the foundation for human health and wellbeing, ecological stability, and economic growth, and can thus be viewed as a cross-cutting dimension of this new development framework.

Box 2.1 Description of SDG 6 on Water Security (UN 2015)

GOAL 6 – Ensure availability and sustainable management of water and sanitation for all

SDG 6 TARGETS

6.1: By 2030, achieve universal and equitable access to safe and affordable drinking water for all

6.2: By 2030, achieve access to adequate and equitable sanitation and hygiene for all and end open defecation, paying special attention to the needs of women and girls and those in vulnerable situations

6.3: By 2030, improve water quality by reducing pollution, eliminating dumping and minimizing release of hazardous chemicals and materials, halving the proportion of untreated wastewater and substantially increasing recycling and safe reuse globally

6.4: By 2030, substantially increase water-use efficiency across all sectors and ensure sustainable withdrawals and supply of freshwater to address water scarcity and substantially reduce the number of people suffering from water scarcity

6.5: By 2030, implement integrated water resources management at all levels, including through transboundary cooperation as appropriate

6.6: By 2020, protect and restore water-related ecosystems, including mountains, forests, wetlands, rivers, aquifers and lakes

6.a: By 2030, expand international cooperation and capacity-building support to developing countries in water- and sanitation-related activities and programmes, including water harvesting, desalination, water efficiency, wastewater treatment, recycling and reuse technologies

6.b: Support and strengthen the participation of local communities in improving water and sanitation management

Another major development in 2015 was the achievement of an agreement to address climate change. The resulting 2015 Paris Agreement was also heralded as a major achievement and success, and sets new flexible goals that are intended to spur innovations in economic development. In focusing on mitigation of greenhouse gases and minimizing climate change, it sets a new target: limit the increase in mean global temperature to 2 °C (Schellnhuber et al. 2016). It also brings attention to the need for effective adaptation to climate change, which is where most of the water management issues (such as extreme weather events, floods, and droughts) intersect with climate change. A more detailed treatment of the ramifications of the 2015 Paris Agreement for water security is provided in Sect. 2.2.2.

A third global dialogue in 2015, the Financing for Development Summit in Addis Ababa, yielded more mixed results. Many found that it did not succeed in identifying visionary solutions for mobilizing new financial resources to implement the new development agenda. The participants of this summit were ostensibly focused mainly on enhancing public resources within developing countries through the use of improved fiscal policies, tax reforms, and reduction in the level of corruption; this approach is based on the understanding that domestic resource mobilization – at US$ 8 trillion a year – far outpaces the "external" funding for development (which includes overseas development aid, private financial flows, and remittances) at US$ 1 trillion (Chhibber 2016). Inclusive and sustainable industrialization was also considered to be central to economic growth, again leaving the onus of transition and formulation of enabling policies on the developing countries (Kjorven et al. 2015). The summit declaration urged governments to incentivize the engagement of the private sector, particularly creating processes that would enhance foreign direct investment (FDI). While the international leadership gathered in Addis Ababa identified the need to build a new forum that would bridge infrastructure gaps in developing countries by mobilizing between US$ 1 and US$ 1.5 trillion, it is not very clear how this mechanism will operate (Chhibber 2016).

A fourth global dialogue took place in Sendai, Japan, with a focus on reducing disaster-related risks. The Third UN World Conference on Disaster Risk Reduction adopted the Sendai Framework, a 15-year, voluntary, non-binding agreement, which aims to achieve "the substantial reduction of disaster risk and losses in lives, livelihoods and health and in the economic, physical, social, cultural and environmental assets of persons, businesses, communities and countries" (UNISDR 2015). This framework urges national governments to assume the primary role in reduction of disaster-related risks, while recognizing that other stakeholders including the private sector must shoulder some of the responsibility. Reading the language of the declaration in which the word "water" appears only once in passing, one can only presume that water-related disasters are implicitly included in the priority areas recognized in the Sendai Framework, even when floods and droughts are not mentioned anywhere in the document.

In addition to the goal- and target-setting within each of these international summit processes, we must also consider their cumulative impact on governmental policies and political perspectives with respect to sustainable development and economic growth. Collectively, there is a need for corresponding seismic-scale shifts in national economic and development planning, while overcoming some serious capacity gaps in most developing countries. An equally important dimension is provision of scientifically developed, evidence-driven policy advice to governments that can help them visualize and achieve short- and long-term success.

2.2.1 Uniqueness of the Development Agenda

Collectively, this 2030 Agenda has been characterized as "transformative"; whether everyone agrees with this characterization remains open to question. Nevertheless, the 2030 Agenda has some unique features.

First, it is universal by design and applies equally to all countries. In doing so, it moves away from the approach of the MDGs which only targeted developing countries. This egalitarian approach is more than superficial and will bring some real consequences in relation to how this agenda is implemented across the world. Based on some early indications, some of the developed countries appear to struggle in coping with the implications of implementing SDGs nationally; to state the obvious, developed countries are not geared towards monitoring development and governance process in quite the same way as these have been undertaken by developing countries as part of the MDG framework from 2000 to 2015. For example, Canada has responded to the 2030 Agenda by acknowledging that SDGs need to play a role in formulating the Federal Sustainable Development Strategy (ECCC 2016); the FSDS outlines how sustainable development targets developed by the Canadian Government in consultation with stakeholders could contribute towards specific SDG targets. This indirect approach of selectively juxtaposing national sustainable development targets with some SDG targets is arguably less effective than the converse approach of defining national development targets based on the framework provided through the 2030 Agenda.

Second, it is comprehensive in that it seeks to completely eliminate almost all of the major social, environmental and developmental problems we face today. Given the inclusive dialogue that preceded the establishment of SDGs, a broad diversity of perspectives and voices have been included in arriving at these aspirations. This diversity is duly reflected in the range of issues that span a dozen or so economic sectors, including but not limited to: health, education, agriculture, fisheries, forestry, energy, water, manufacturing, urban infrastructure, transport, shipping, and tourism. An integrated and collective approach would be essential to ensure that these sectors do not work at cross-purposes as they attempt to meet the corresponding SDG targets.

Third, as a result of its comprehensiveness, the SDGs are quite complex as shown by the large number of goals and targets. One may argue that the political negotiation process that led to this current formulation did not make a sufficient attempt to better consolidate goals into fewer, and more cohesive goals and targets. Some alternative approaches, such as that adopted by Dr. Jeffrey Sachs, presented much more compact sets of goals (SDSN 2014). This complexity *inter alia* poses a challenge for monitoring progress during the 15-year period, most notably for developing countries which are typically over-burdened with reporting requirements and chronically lack the capacity – in terms of human and financial resources – to provide adequate, evidence-based reports. By the same token, achieving cohesiveness across the various economic sectors could also be problematic.

Fourth, the SDGs are quite ambitious because they aim to fix the major global problems in a 15-year span. Reviewing the language used and the underlying intention in shaping the goals and targets, it is obvious that the global community aims for complete or near-complete eradication of major problems. Learning from the MDG experience, in which governmental actions typically gravitated towards easy-to-solve challenges while the more difficult ones languished (UNU and UNOSD 2013), some governments may again attempt to adopt a similar preferential approach. Were this to happen, we will encounter yet again the inability to address the problems encountered by the most vulnerable populations in remote and/or rural settings (UN-Water 2010). Substantial and evidence-based analysis of the implementation timelines does not exist for most of the SDG targets, leaving one to speculate whether 15 years is too short a period or too long (Schuster-Wallace and Sandford 2015).

2.2.2 Paris Agreement and Water Security

As discussed earlier, after some 20 years of ongoing negotiations, world leaders gathered in Paris in December 2015 to sign off on the landmark Paris Agreement. This ostensible success was driven by the clear and present need for action, feasibility of achieving targets through political will and the use of available technology, and the simplicity of agreeing to an easily describable target of restricting the global average temperature increase to 2 °C above pre-industrial levels (Schellnhuber et al. 2016). A key action area of the Paris Agreement is to document national actions through Nationally Determined Contributions, or NDCs; much work still needs to be done to develop common metrics through which progress on NDCs can be monitored while ensuring that these contributions are indeed incremental over pre-existing initiatives. It is laudable that the parties to the United Nations Framework Convention on Climate Change (UNFCCC) have agreed to revisit the targets in view of progress every 5 years (Bodansky 2016).

It is well established through the work published by the Intergovernmental Panel on Climate Change (IPCC) that many regions of the world will experience altered hydrological systems that affect the quality and quantity of water resources (IPCC 2014), and yet adaptation to this new hydrological regime has largely remained a distant issue in the climate change dialogue (Bodansky 2016). IPCC reports and preponderance of scientific evidence demonstrates that the sharpest societal impact of climate change takes the form of drastic changes to the hydrological cycle: extreme weather events, record-breaking floods, long and extensive droughts, sea-level rise, and changes to the extent of cryosphere (read: melting glaciers and polar caps). These water-related impacts are felt across the world and without distinction to the state of economic, social, or political development of a country or a region. The Paris Agreement offers some course correction by paying attention to adaptation (e.g., seeking balance between mitigation and adaptation in Article 9.4) and reviewing institutional and capacity issues. However, it still largely views adaptation as a developing-country problem. In this respect, Laukkonen notes that "it is not sufficient to concentrate on either mitigation or adaptation, but a combination of these results in the most sustainable outcomes. Yet, these two strategies do not always complement each other, but can be counterproductive" (Laukkonen et al. 2009). It stands to reason that investments for adaptation to climate change must primarily address climate variability and in some cases drastic changes to hydrology in order to achieve water security and human development (UN-Water 2010).

In paragraph 54 of the Paris Agreement, Parties have agreed to mobilize US$ 100 billion per year through to the year 2025 when they will set a new collective goal. When viewed together with Article 9.4, one may argue that a significant component of these mobilized funds should be dedicated towards achieving water security. Whether such mobilization of resources takes place and whether a significant fraction (say, close to half) is dedicated to adaptation-related efforts, however, remains to be seen.

2.3 Intersection of Sustainable Development and Water Security

2.3.1 Correlation Between Social Marginalization and Water Insecurity

Previous published research has pointed to direct links between health/wellbeing indicators and access to safe drinking water and adequate sanitation (UNU and UNOSD 2013; UNICEF and WHO 2015). One can take the argument even further: water insecurity is a result of social marginalization and can further exacerbate it. Data from developing countries shows that deprivation of water security links also with populations that are socially and politically marginalized; these include people living in urban slums and peri-urban areas as well as rural populations (UNU and UNOSD 2013; UN-Water 2014). The UN-Water Global Analysis and Assessment of Sanitation and Drinking-Water (GLAAS) report suggests that there is a strong bias in lack of water and sanitation service provisioning based on income levels (UN-Water 2014). For populations that are socially and politically marginalized, governments may not prioritize investment of resources to build and manage water/sanitation delivery infrastructure. Conversely, water insecurity ensures that people living in slums or remote rural areas remain trapped in a vicious downward spiral of poverty and declining wellbeing.

2.3.2 Human Rights as a Driving Force for Development?

The United Nations General Assembly declared water and sanitation to be fundamental human rights in 2010 (Resolution A/RES/64/292); this was followed a few weeks later by a similar but more detailed resolution by the UN Human Rights Council (Resolution 15/9). This landmark decision by UN Member States was viewed by many as a major achievement and a logical outcome of many decades of lobbying that water and sanitation should be recognized as a human right. In the years since these resolutions were passed, much has been written about the pros and cons of declaring these services as a human right; many countries have included these rights in their constitution, while others still struggle with the ramifications of these rights in a legal and legislative context (Boyd 2012; Chociej and Adeel 2012).

There is also a minority of scholars who argue that adoption of these services as a human right may actually hinder their provisioning to the under-privileged (Pardy 2012). That is based on two premises: First, the general public in common discourse misconstrues these human rights as akin to free delivery of water and sanitation to everyone. In a pragmatic way, delivery of any public service requires financial resources that can come from the taxpayer, or through tariffs imposed on consumers, or through "foreign" transfers from various development partners. There really is no free lunch when it comes to provisioning of water and sanitation services. And yet, incorrect public perceptions lead to heated, politicized, and polarized debates on water pricing in public and media circles. Second, in water-scarce environments there may be undue burden imposed on limited water resources, and traditional consumers of water such as the agricultural sector might suffer. Calculating the tradeoff between competing priorities could likely become a politicized process, in which those who are politically marginalized may be further disenfranchised from their rights.

Despite the foregoing criticism, the legal and ethical pressures for achieving water security on governments and the collective international community must be sustained. One might, therefore, envisage a more optimistic scenario: Governments and various related stakeholders take the achievement of the human right to water seriously and invest significant resources towards its success; the SDGs already provide a framework for such policies and management approaches.

2.4 Roadblocks to Achieving Water Security

2.4.1 Insufficient Resource Mobilization

Some early estimates are available of the scale and scope of investments needed to achieve the SDGs. Based on the available data, it is estimated that somewhere between 1.8 and 2.5% of the global GDP will need to be invested in order to achieve all the water related SDGs alone (UNU and UNOSD 2013). This amounts to a very significant increase in financial resources invested into infrastructure development, sustained management, and capacity development. In a recent study that analyzed ten developed and developing countries, it has been demonstrated that the ambitious development agendas are not matched by sufficient financial resources (Schuster-Wallace and Sandford 2015). Most developing countries recognize this shortfall.

It remains an open question whether governments have the political will and the financial wherewithal to make new and incremental investments in order to achieve water security. This is an area in which the private sector can play a role in mobilizing the needed financial resources – irrespective of whether these resources are 'domestic' or mobilized through FDI (Chhibber 2016). Two crucial criteria for private sector involvement must be met. First, engagement of the private sector, linked obviously to turning a profit, must be balanced against monitoring and oversight by the public sector. Regulations and monitoring are critical to ensuring that the public service is protected and is made available to socially and economically marginalized communities. Second, innovative business models are needed to serve those who are economically disadvantaged. The mobile phone industry offers an interesting model in which companies have continued to turn a profit even when serving the 'base of the pyramid.' As an example, there are over 250 million people in India who have access to mobile phones but not toilets; this situation points to the notion that there likely is affordability and willingness to pay for water and sanitation services.

2.4.2 *Overcoming Capacity Gaps*

Prior experience in capacity development demonstrates that success can be achieved only when multiple dimensions are considered in a cohesive and integrated manner (Franks 1999). Four dimensions of capacity gaps have been identified in the context of water-related initiatives: insufficient numbers of and inadequately trained human resources; lack of appropriate and affordable technologies for delivery of water and sanitation services; lack of institutional, regulatory, and governance mechanisms for effective management and delivery of services; and, inadequate financial resources. Frequently, development projects and initiatives that aim to achieve water security either ignore the need for capacity development, or do so without addressing all four elements (Adeel et al. 2015). Training of personnel, in the absence of reasonable financing, supporting technology, and enabling institutional arrangements leads to sub-optimal outcomes. This means that capacity building needs to be nested within broader institutional reform at a national scale. An anecdotal analysis of ten countries reveals that building up of these capacities is generally not keeping pace with the growing demands due to increased population, changing lifestyles and consumption patterns, competing water usage by agriculture and industrial sectors, and increased economic activity.

With a set of well-defined targets to be met by 2030, concerted and integrated efforts for capacity development are needed; this will require multi-year, extended engagement of governments, donors, and other stakeholders (Reed 2012). The United Nations system has a well-defined mandate to develop the capacity of its Member States, and it happens to be the only organization that has a footprint and local presence in all developing countries (Adeel et al. 2015). It needs to revert to its approach of 'Delivering as One' and enhance integration of capacity development activities undertaken by various UN organizations and agencies.

Within developing countries, creation of national and regional institutions that can support such capacity development initiatives is a must (Adeel et al. 2015). The international community has a crucial role to play in filling the capacity gaps. Success of the international community, in turn, is tied to enabling conditions provided by national and sub-national governments (Adeel et al. 2015).

2.5 Framing Enabling Policies

2.5.1 Priority Setting by National Governments

Successful achievement of the SDGs will require first and foremost a clear expression of support and a commitment to re-tool national development priorities. Given the broad spectrum of goals incorporated into the SDG framework, it is essential that national development planning account for the SDGs and establish monitoring programs that incorporate specific and time-bound targets (Schuster-Wallace and Sandford 2015). This integrated approach for development planning should preferably be executed through an apex governmental mechanism with an implementation and monitoring mandate. Sufficient evidence regarding current state and development trajectory must be weighed in when planning for management of water resources and delivery of water-related services. Competing demands for water must be resolved through engaged stakeholder dialogues, as no pre-packaged solution will fit all situations.

Reporting on progress for implementation of SDGs is crucial, must be centralized for each country and linked to national level monitoring programs. Transparency in reporting means greater accountability and supports sustained political focus on achieving the SDG targets (Schuster-Wallace and Sandford 2015). This accountability should, by default, also similarly extend to initiatives supported by the international community. National reports on monitoring and sharing of water resources would help achieve a balance in supply and demand of water. Water use efficiency by various sectors, most particularly agriculture, would also be made obvious through such reporting.

2.5.2 *Bringing Innovation to Resource Mobilization*

As noted earlier, the financial resources required for achieving global water security adds up to an astounding amount of US $1.4 to US $2.6 trillion a year, roughly corresponding to 1.8–2.5% of the global GDP (UNU and UNOSD 2013). A rough estimate indicates that such investments would leverage at least US$3 trillion a year in benefits that would accrue in perpetuity. This argument for cost-benefit ratio on its own is not sufficient, however, and business as usual will not result in mobilizing the requisite financial resources. The current water "sector" mobilization stands at about US $530 billion a year on worldwide basis (WWC 2015), requiring at least another trillion dollars in financial capital for meeting the water-related SDGs; clearly some out of the box and innovative thinking is required. The following four innovative approaches need to be further explored.

First, national development agendas need to be re-structured in a way that allocation of budgetary resources is based on planning across sectors. Most notable is the need to align planning along the water, food, and energy security nexus. In principle, re-prioritizing of national development would result in optimization of financial resource allocation. Second, the financial resources allocated under the 2015 Paris Agreement could be utilized for water-related adaptation initiatives, as discussed earlier. Third, the private sector could be incentivized to invest in projects that address SDG 6 targets; the private sector can bring in not only the financial capital but also the business management approaches needed for logarithmic growth in service provisioning and the technological know-how for implementing effective on-the-ground solutions (Adeel 2014). To put matters in perspective, the capitalization of the international financial markets stands at over US $60 trillion a year and can be a valuable resource for funding (World Bank 2016). Fourth, corruption and graft in the water sector is estimated to consume 30% of funding resources (Transparency International 2008). Therefore, eliminating corruption and graft from the water sector can unlock sizeable volume of resources; doing so would also strengthen the arguments for attracting new and hitherto untapped investments into the water domain.

2.5.3 *Role of the "International Community"*

A significant engagement of the international community can facilitate the achievement of SDGs; this can take the form of development assistance, loans, grants and foreign direct investment. The international community can and should also play a central role in building capacity of developing countries to understand, manage, and respond to their own problems. Having said that, there is a growing literature that documents the ineffectiveness of international assistance; for example, Chan and Chung (2015) have placed the blame for countries' inability to achieve many of the Millennium Development Goals (MDGs) by 2015 on failure of development assistance. There are some specific trends in international assistance that can be highlighted in this context: (a) the international assistance, particularly coming through bilateral aid agencies, is tied to the technical and technological support coming exclusively from the corresponding donor countries, thereby often reducing its effectiveness (Mosley et al. 2004); (b) the aid programs tend to bypass the governmental channels, ostensibly to short-circuit bureaucracy and corruption, and end up tying them to political and geostrategic agendas (Kelegama 2012); (c) the international assistance programs generally tend to shy away from long-term investments in institutional and governance mechanisms (Kelegama 2012); and, (d) many assistance programs impose criteria and conditions on recipients that even donor countries cannot meet (Hout 2002). The analysis undertaken by many researchers implies that the international community needs a renewed effort to meet the principles outlined in the 2005 Paris Declaration for Aid Effectiveness: ownership, alignment, harmonization, managing for results, and mutual accountability (OECD 2006; Michaelowa and Weber 2007; Ravenborg et al. 2013; Chan and Chung 2015).

Further, there are new donors emerging that operate outside the remit of OECD, and hence the 2005 Paris Declaration; these include countries like China, Malaysia, and India. There is an ostensible trend of these investors focusing much more on geostrategic alliances and government-based initiatives, as outlined in the 'Beijing Consensus' (Yao 2011). A number of large-scale energy and transport initiatives are examples of such alliances: the 1000 Electricity Transmission and Trade Project for Central Asia and South Asia (CASA 1000) between Kyrgyz Republic, Tajikistan, Afghanistan and Pakistan; the Turkmenistan-Afghanistan-Pakistan-India Pipeline (TAPI); and, the China Pakistan Economic Corridor (CPEC). Despite anecdotal collaboration of these new development partners with the OECD-based organizations, there remains a strong need for reverting to and emphasizing the Paris Declaration percepts of neutrality, good governance, and mutual cooperation. While national and sub-national governments will need to carry the major burden of planning and resourcing for implementation of SDGs, the international community can intervene and help expedite the time horizon needed for gearing up institutional, financial, human, and technological resources.

2.6 Conclusion: Outlook for Success of the SDGs

The foregoing dialogue shows how water security, defined broadly, intersects with the emerging and transformative international development agenda. Water occupies a pivotal position – as both a resource and a service sector – for achieving the major objectives related to poverty, health, education, hunger, gender equity, climate change, and sustainability within this development framework. Water security also intersects horizontally across the global dialogues on climate change, disasters, and financing for development. There has been an increasing international recognition of this importance of water security, most notably in the policy framing for numerous governments. The world business community has also given due recognition to water security for a number of years; for example, water insecurity has consistently emerged as a top-five threat in terms of impact and likelihood in the global risk landscape developed at the Davos World Economic Forum (WEF 2016).

Despite all the political and economic parameters pointing to the significance of water security, it has not been framed as such in most national development planning processes. The information provided in this chapter supports the argument that the 2030 Agenda, coupled with the 2015 Paris Agreement, offers an opportunity to national governments as well as the international development community to hit the 'reset' button. Such resetting of national development plans will be hampered by two significant roadblocks described earlier: financial mobilization and capacity gaps. With some innovative thinking and creation of new partnerships, both of these roadblocks can be surmounted. The real challenge for governments is development of integrated national development narratives, creation of harmonized and cohesive legislative and legal frameworks, leading stakeholder conversations, and enhancement of accountability through monitoring systems. It stands to reason that national realities in terms of political, cultural, and societal settings will play an instrumental role in devising successful strategies and policy approaches.

There is another area of importance, but outside the scope of this chapter: actions that takes place at the sub-national level. For example, municipal-level responses will be central to national policies, given rapid urbanization and growth of numerous megacities (with populations of over ten million) around the world. The municipal governments increasingly exercise tremendous political and economic influence, and often become drivers of national development narratives. Similarly, some actions at provincial or state-level jurisdictions would also make sense; such coalitions have already emerged in North America around the issues related to climate change.

It is important to note that the research community and academia at large have a major role to play in investigating new ideas for implementation around the 2030 Agenda. The greatest opportunity lies in exploring new business models that allow provisioning of water services and management of resources, particularly in developing country settings. Developing and field-testing these business models would in turn require new collaborations between the private sector and academia; perhaps the United Nations system can serve as the facilitator and moderator between these two groups.

Annex 2.1: The Millennium Development Goals

Goal 1: Eradicate extreme poverty and hunger

- Target 1A: Halve, between 1990 and 2015, the proportion of people living on less than $1.25 a day
- Target 1B: Achieve Decent Employment for Women, Men, and Young People
- Target 1C: Halve, between 1990 and 2015, the proportion of people who suffer from hunger

Goal 2: Achieve universal primary education

- Target 2A: By 2015, all children can complete a full course of primary schooling, girls and boys

Goal 3: Promote gender equality and empower women

- Target 3A: Eliminate gender disparity in primary and secondary education preferably by 2005, and at all levels by 2015

Goal 4: Reduce child mortality rates

- Target 4A: Reduce by two-thirds, between 1990 and 2015, the under-five mortality rate

Goal 5: Improve maternal health

- Target 5A: Reduce by three quarters, between 1990 and 2015, the maternal mortality ratio
- Target 5B: Achieve, by 2015, universal access to reproductive health

Goal 6: Combat HIV/AIDS, malaria, and other diseases

- Target 6A: Have halted by 2015 and begun to reverse the spread of HIV/AIDS
- Target 6B: Achieve, by 2010, universal access to treatment for HIV/AIDS for all those who need it
- Target 6C: Have halted by 2015 and begun to reverse the incidence of malaria and other major diseases

Goal 7: Ensure environmental sustainability

- Target 7A: Integrate the principles of sustainable development into country policies and programs; reverse loss of environmental resources
- Target 7B: Reduce biodiversity loss, achieving, by 2010, a significant reduction in the rate of loss
- Target 7C: Halve, by 2015, the proportion of the population without sustainable access to safe drinking water and basic sanitation
- Target 7D: By 2020, to have achieved a significant improvement in the lives of at least 100 million slum-dwellers

Goal 8: Develop a global partnership for development

- Target 8A: Develop further an open, rule-based, predictable, non-discriminatory trading and financial system
- Target 8B: Address the Special Needs of the Least Developed Countries (LDCs)
- Target 8C: Address the special needs of landlocked developing countries and small island developing States
- Target 8D: Deal comprehensively with the debt problems of developing countries through national and international measures in order to make debt sustainable in the long term

- Target 8E: In co-operation with pharmaceutical companies, provide access to affordable, essential drugs in developing countries
- Target 8F: In co-operation with the private sector, make available the benefits of new technologies, especially information and communications

References

Adeel, Z. (2012). A human development approach to water security. In H. Bigas, T. Morris, R. Sandford, & Z. Adeel (Eds.), *The global water crisis: Addressing an urgent security issue*. Hamilton: United Nations University.

Adeel, Z. (2014). Engaging and incentivizing the private sector: An emerging opportunity for the water world. *New Water Policy and Practice Journal, 1*(1), 20–25.

Adeel, Z., Aslov, S., Maestu, J., & Unver, O. (2015). *Water cooperation – Views on progress and the way forward*. Hamilton: United Nations University.

Bodansky, D. (2016). The Paris climate change agreement: An new hope. *American Journal of International Law, 110*. Available at SSRN: https://ssrn.com/abstract=2773895.

Boyd, D. (2012). The right to water: Moving from international recognition to national action. In H. Bigas, T. Morris, R. Sandford, & Z. Adeel (Eds.), *The global water crisis: Addressing an urgent security issue*. Hamilton: United Nations University.

Chan, S.-J., & Chung, Y.-H. (2015). Trends and challenges of aid effectiveness: The rise of Asia. In I.-H. Cheng & S.-J. Chan (Eds.), *International education aid in developing Asia*. Singapore: Springer.

Chhibber, A. (2016). *Assessing and evaluating the Addis Ababa Action Agenda (AAAA)*. NIPFP working paper series No. 166. New Delhi: National Institute of Public Finance and Policy (NIPFP).

Chociej, Z., & Adeel, Z. (2012). Legal and ethical dimensions of a right to water. In H. Bigas, T. Morris, R. Sandford, & Z. Adeel (Eds.), *The global water crisis: Addressing an urgent security issue*. Hamilton: United Nations University.

ECCC. (2016). *Planning for a sustainable future: A federal sustainable development strategy for Canada 2016–2019*. Gatineau: Environment and Climate Change Canada (ECCC).

Fehling, M., Nelson, B. D., & Venkatapuram, S. (2013). Limitations of the millennium development goals: A literature review. *Global Public Health, 8*(10), 1109–1122.

Franks, T. (1999). Capacity building and institutional development: Reflections on water. *Public Administration and Development, 19*, 51–61.

Fukuda-Parr, S., Yamin, A. E., & Greenstein, J. (2014). The power of numbers: A critical review of millennium development goal targets for human development and human rights. *Journal of Human Development and Capabilities, 15*(2–3), 105–117.

Hout, W. (2002, June). Good governance and aid: Selectivity criteria in development assistance. *Development and Change, 33*(3), 511–528.

IPCC. (2014). Summary for policymakers. In C. B. Field, V. R. Barros, D. J. Dokken, K. J. Mach, M. D. Mastrandrea, T. E. Bilir, M. Chatterjee, K. L. Ebi, Y. O. Estrada, R. C. Genova, B. Girma, E. S. Kissel, A. N. Levy, S. MacCracken, P. R. Mastrandrea, & L. L. White (Eds.), *Climate change 2014: Impacts, adaptation, and vulnerability. Part A: Global and sectoral aspects. Contribution of working group II to the fifth assessment report of the intergovernmental panel on climate change* (pp. 1–32). Cambridge/New York: Cambridge University Press.

IRF2015. (2013). *Post-2015: Framing a new approach to substantial development*. Washington DC: Independent Research Forum 2015 (IRF2015). Retrieved April 13, 2016, from http://pubs.iied.org/G03559.html?k=framing%20a%20new%20approach

Kelegama, S. (Ed.). (2012). *Foreign aid in South Asia: The emerging scenario*. New Delhi: Sage.

Kjorven, O., Nippita, N., & Wietzke, F. B., (2015). *The Addis Ababa Action Agenda: Strengths, weaknesses, and the way ahead.* New York: UNICEF Connect, United Nations Children's Fund (UNICEF). Retrieved August 22, 2016, from https://blogs.unicef.org/blog/the-addis-ababa-action-agenda-strengths-weaknesses-and-the-way-ahead/

Laukkonen, J., Blanco, P. K., Lenhart, J., Keiner, M., Cavric, B., & Kinuthia-Njenga, C. (2009). Combining climate change adaptation and mitigation measures at the local level. *Habitat International, 33*, 287–292.

MA. (2005). *Ecosystems and human well-being: Synthesis.* Washington, DC: Millennium Ecosystem Assessment (MA)/Island Press.

Michaelowa, K., & Weber, A. (2007). *Aid effectiveness in Primary, secondary and tertiary education.* Paris: UNESCO. Retrieved December 23, 2015, from http://unesdoc.unesco.org/images/0015/001555/155559e.pdf

Mosley, P., Hudson, J., & Verschoor, A. (2004, June). Aid, poverty reduction and the 'new conditionality'. *The Economic Journal, 114*, F217–F243.

OECD. (2006). *Harmonising donor practices for effective aid delivery, volume 2 – Budget support, sector wide approaches and capacity development in public financial management, Organisation for Economic Co-Operation and Development (OECD), DAC guidelines and reference series.* Paris: OECD.

Pardy, B. (2012). False panacea: The human right to water. In H. Bigas, T. Morris, R. Sandford, & Z. Adeel (Eds.), *The global water crisis: Addressing an urgent security issue.* Hamilton: United Nations University.

Ravnborg, H. M., Larsen, R. K., Vilsen, J. L., & Funder, M. (2013). *Environmental governance and development cooperation achievements and challenges* (p. 15). Copenhagen: Danish Institute for International Studies – DIIS Report.

Reed, L. K. (2012). Capacity building as a policy instrument in water conservation: A case study on commercial, industrial, and institutional consumers. *Water Resources Management, 26*(13), 3819–3829.

Schellnhuber, H. J., Rahmstorf, S., & Winkelmann, R. (2016). Why the right climate target was agreed in Paris. *Nature Climate Change, 6*, 649–653.

Schuster-Wallace, C. J., & Sandford, R. (2015). *Water in the world we want.* Hamilton: United Nations University Institute for Water, Environment and Health and United Nations Office for Sustainable Development, UNU-INWEH.

SDSN. (2014). *Framing sustainable development goals targets and indicators, Issue brief.* New York: Sustainable Development Solutions Network (SDSN).

Transparency International. (2008). Global corruption report 2008 – Corruption in the water sector. Cambridge: Cambridge University Press. Retrieved February 18, 2016, from http://issuu.com/transparencyinternational/docs/global_corruption_report_2008?e=2496456/2011923

UN. (2015). *Transforming our world: The 2030 agenda for sustainable development, A/RES/70/1.* New York: United Nations.

UNDP. (1994). *Human development report, 1994. Cultural liberty in today's diverse world.* New York: United Nations Development Programme (UNDP).

UNICEF and WHO. (2015). *Progress on sanitation and drinking water – 2015 update and MDG assessment.* New York: United Nations Children's Fund (UNICEF) and World Health Organization (WHO).

UNISDR. (2015). *Sendai framework for disaster risk reduction 2015–2030.* Geneva: The United Nations Office for Disaster Risk Reduction (UNISDR).

UNSGAB. (2015). The UNSGAB journey: United Nations secretary-general's advisory board on water and sanitation (2004–2015). New York: UNSGAB. Retrieved January 15, 2016, from https://sustainabledevelopment.un.org/topics/water/unsgab

UNU and UNOSD. (2013). *Water for sustainability: Framing water within the post-2015 development Agenda.* Hamilton: United Nations University Institute for Water, Environment and Health (UNU-INWEH), UN Office of Sustainable Development (UNOSD) and Stockholm Environment Institute (SEI), UNU-INWEH.

UN-Water. (2010). *Climate change adaptation: The pivotal role of water*. Geneva: UN-Water. Retrieved August 18, 2016, from http://www.unwater.org/downloads/unw_ccpol_web.pdf

UN-Water. (2013). *Water security & the global water agenda, a UN-water analytical brief*. Hamilton: United Nations University.

UN-Water. (2014). *Investing in water and sanitation: Increasing access, reducing inequalities, UN-Water global analysis and assessment of sanitation and drinking-water, GLAAS 2014 report*. Geneva: World Health Organization.

WEF. (2016). *The global risks report 2016, 11th edition*. Davos: World Economic Forum (WEF).

World Bank. (2016). *Market capitalization of listed domestic companies (current US$), World Federation of Exchanges database*. Washington, DC: World Bank. Retrieved August 22, 2016, from http://data.worldbank.org/indicator/CM.MKT.LCAP.CD

World Bank and IMF. (2005). *Synthesis – 2005 review of the PRS approach: Balancing accountabilities and scaling up results*. Washington, DC: The World Bank and The International Monetary Fund (IMF).

WWC. (2015). *Water: Fit to finance? Catalyzing national growth through investment in water security, Report of the high level panel on financing infrastructure for a water-secure world, April 2015, World Water Council (WWC)*. Marseilles: World Water Council (WWC).

Yao, Y. (2011). *Beijing consensus or Washington consensus: What explains China's economic success*. Washington, DC: World Bank. Retrieved December 23, 2015, from https://openknowledge.worldbank.org/handle/10986/6098

Chapter 3
Water, Law and Equity

Owen McIntyre

Abstract While the legal concept of "equity" as a body of rules is routinely employed in a number of fields of international law relevant to the management of water resources and the supply of water services, including international environmental law and international human rights law, it is absolutely central to international water resources law, *i.e.* the body of rules concerned with the inter-State allocation of rights in the uses and benefits of shared transboundary waters. As suggested by the cardinal principle of international water law, the principle of equitable and reasonable utilisation, the overarching objective of this body of law is to determine how such rights in shared water resources can be allocated equitably, taking account of a range of relevant factors and considerations. Of course, the ultimate aim of the corpus of international rules applying to water resources management is that of ensuring human well-being by achieving the greatest attainable measure of "water security". However, there exists little consensus about the normative nature or content of the equitable rules or principles to be applied in this context or about their legal implications for the cooperative management of transboundary waters. This chapter attempts to map the use of equitable concepts in cognate areas of international law in order to shed some light on its possible application in the field of international water law.

Keywords International water law • Equity • Proportionality • Equitable and reasonable utilisation

3.1 Introduction

The rather vague language of 'equity' pervades the various bodies of rules of international law which apply to the management, protection, utilisation, allocation and supply of freshwater resources, including international environmental law, international human rights law and of course international water resources law. While this inevitably results in considerable uncertainty regarding the precise normative

O. McIntyre (✉)
School of Law, University College Cork, Cork, Ireland
e-mail: o.mcintyre@ucc.ie

D. Devlaeminck et al. (eds.), *The Human Face of Water Security*, Water Security in a New World, DOI 10.1007/978-3-319-50161-1_3

implications of such rules, a measure of normative vagueness permits the flexibility required to secure the participation of hesitant State actors in the ongoing gradual and progressive elaboration of the kind of sophisticated legal regimes necessary for the cooperative and sustainable management of an increasingly scarce and contested resource. It should also be remembered that no two water systems, comprising either international watercourses and/or aquifers or water services supply systems, are remotely alike – hydrologically, ecologically, climatically, socially, economically, demographically or politically – and so a measure of flexibility is welcome in the applicable rules and standards in order to facilitate taking account of relevant contextual factors. It must be possible, however, notwithstanding the need for vagueness and flexibility, to divine certain core elements and values inherent to the concept of equity, as least insofar as it applies to water resources and services.

Equity is also central to achieving the overarching objective of "water security", which has recently been defined as

> the capacity of a population to safeguard sustainable access to adequate quantities of acceptable quality water for sustaining livelihoods, human well-being, and socio-economic development, for ensuring protection against water-borne pollution and water-related disasters, and for preserving ecosystems in a climate of peace and political stability (UN-INWEH 2013, p. 1).

Thus, the values inherent to the goal of water security – the access of individuals to water, the socio-economic development of States, the prevention of water pollution and preservation of ecosystems – correspond very closely to those pursued by international water law, and by related norms of international environmental law and international human rights law, which are in each case informed by the pervasive concept of equity.

3.2 Equity in International Law

3.2.1 International Environmental Law

International environmental law has long relied on the concept of equity in high-profile declaratory and conventional instruments including, for example, Articles 3(1) and 4(2)(a) of the 1992 United Nations Framework Convention on Climate Change (UNFCCC 1992), and Articles 1 and 15(7) of the 1992 Convention on Biological Diversity (CBD 1992), both of which instruments have a role in informing normative frameworks for water resources management as regards climate change adaptation and the maintenance of aquatic and water-related ecosystems. As early as 1978 the United Nations Environmental Programme (UNEP 1978) facilitated the adoption of a seminal set of draft principles to guide States in the environmental protection and cooperative utilisation of shared natural resources, including water resources, which declared equity to be the key requirement 'with a view to controlling, preventing, reducing or eliminating adverse environmental effects which may result from the utilization of such resources' (UNEP 1978). More generally, the concept of inter-generational equity articulated in Principle 3 of the Rio Declaration (UNCED 1992),[1] and the notion of intra-generational equity encapsulated in the Principle of 'common but differentiated responsibilities' set out under Rio Principle 7[2] illustrate the absolutely central role of the concept of equity to the principle of sustainable development, the key organising principle underlying all norms of modern international environmental law.[3] This has prompted leading commentators to note that 'in many respects, UNCED was about equity' largely because 'in the absence of detailed rules, equity can provide a conveniently flexible means of leaving the extent of rights and obligations to be decided at a subsequent date' (Sands and Peel 2012, pp. 213–214). Through its articulation in Article 5(c) of the 1992 UNECE Water Convention, the principle of inter-generational equity has been expressly endorsed in one of only two global instruments applying to shared natural resources (UNECE 1992), and this author has noted that 'the utilisation, for example, of groundwaters in a manner or at a rate inconsistent with their natural capacity to be replenished would clearly prejudice the right of future generations to enjoy the use of such waters, and the references in the 1997 [UN Watercourses] Convention to the goal of sustainable utilisation are intended to mitigate against such utilisation' (McIntyre 2007, p. 260). Of course the principle of equitable and reasonable utilisation, the cardinal and overarching principle of international water law, is now widely regarded as very closely linked to sustainable development in that it 'provides, indeed requires, that states take into consideration the factors tied to sustainable development of the resource, thus providing the legal framework for operationalising this concept' (Wouters and Rieu-Clarke 2001, p. 283; Kroes 1997, p. 83; McIntyre 2007, p. 247). Clearly, this linkage compounds the relevance of the environmental law principles of inter-generational and intra-generational equity to the application of international water law.

However, while both principles are concerned with achieving some form of equitable distribution of costs and benefits in the use of environmental resources, they serve to identify the parties to whom equitable considerations should apply rather than to inform the normative meaning of the notion of equity as it applies in the field of international environmental law (McIntyre 2013a, p. 114). Commentators are

---

[1] Principle 3 of the 1992 Rio Declaration on Environment and Development provides:

The right to development must be fulfilled so as to equitably meet developmental and environmental needs of present and future generations.

[2] Principle 7 of the 1992 Rio Declaration on Environment and Development provides:

States shall cooperate in a spirit of global partnership to conserve, protect and restore the health and integrity of the Earth's ecosystem. In view of the different contributions to global environmental degradation, States have common but differentiated responsibilities.

The developed countries acknowledge the responsibility that they bear in the international pursuit of sustainable development in view of the pressures their societies place on the global environment and of the technologies and financial resources they command.

[3] In 1987 the Brundtland Commission provided a landmark definition of "sustainable development" as:

Development that meets the needs of the present without compromising the ability of future generations to meet their own needs.

ready to point out that, though 'many environmental treaties refer to or incorporate equity or equitable principles … treaties rarely provide a working definition of equity' (Sands and Peel 2012, p. 119). Incorporation of the rather nebulous concept of equity even clouds attempts to codify and articulate core, foundational rules of international environmental law, such as the general duty of States to prevent significant transboundary harm. For example, Articles 9 and 10 of the International Law Commission's (ILC) 2001 Draft Articles on the Prevention of Transboundary Harm from Hazardous Activities sought to require States to seek 'acceptable solutions regarding measures to be adopted in order to prevent significant transboundary harm … based on an equitable balance of interests' (ILC 2001). As with leading textual formulations of the international water law principle of equitable and reasonable utilisation, the 2001 ILC Draft Articles merely provide an indicative list of factors relevant to achieving such solutions.

3.2.2 International Human Rights Law

The language of equity is equally ubiquitous, though nonetheless uncertain, in the international discourse on the human right to water and sanitation. Notably, General Comment No. 15, the seminal document adopted in 2002 by the UN Committee on Economic, Social and Cultural Rights (CESCR) setting out the legal origins and normative content in international law of the purported human right to water, alludes to the concept in several contexts. For example, in relation to the need to ensure access to water resources for agriculture in order to realise the right to adequate food, paragraph 7 calls upon State parties to ensure 'that disadvantaged and marginalized farmers, including women farmers, have *equitable* access to water and water management systems' (CESCR 2002). Thus equity is here associated with the obligation of State parties to avoid any form of discrimination. In turn, paragraph 27, which concerns the obligation of States to ensure that water services are affordable, provides that:

> Any payment for water services has to be based on the principle of *equity*, ensuring that these services, whether privately or publicly provided, are affordable for all, including socially disadvantaged groups. Equity demands that poorer households should not be disproportionality burdened with water expenses as compared to richer households (CESCR 2002).

Here the Committee appears to employ a form of distributive equity based on the concept of proportionality, or at least on the avoidance of disproportionate cost for the poor.

More recently, the targets set out under Sustainable Development Goal 6, which commits the international community to 'ensure availability and sustainable management of water and sanitation for all', and is very closely linked to realisation of the human rights to water and sanitation, place considerable reliance on the notion of equity. Recent methodological guidance produced by UN-Water to inform monitoring of the SDG targets on drinking water and sanitation, explains '*equitable* access to safe and affordable drinking water for all' under Target 6.1 and '*equitable* sanitation and hygiene for all' under Target 6.2 in terms of 'progressive reduction and elimination of inequalities between population sub-groups' (UN-Water 2015). Thus, it remains unclear whether the equitable values inherent to the obligations imposed under the purported human rights to water and sanitation merely require the progressive elimination of discriminatory practices regarding access to drinking water and adequate sanitation or the urgent redistribution of resources in order to eliminate all and any inequality of access to services.

3.2.3 International Water Resources Law

It is in relation to the body of international law relating to the use, management and protection of shared international freshwater resources, however, that the concept of equity plays an absolutely central role. The principle of equitable and reasonable utilisation, as now enshrined in Articles 5 and 6 of the 1997 UN Watercourses Convention (UNWC 1997), is universally understood to be the cardinal and overarching rule in the field (ILC 1994, p. 222). As the name suggests, equity is the key normative value informing application of this fundamental legal principle, as it entitles each co-basin State to an equitable and reasonable use of transboundary waters flowing through its territory, or of the benefits deriving therefrom. In essence, the principle of equitable and reasonable utilisation involves the allocation of rights in the uses and benefits of shared water resources on the basis of a distributive conception of equity having regard to all relevant factors. This suggests that uses and benefits will be shared in proportion to each basin State's needs (McIntyre 2007, p. 147–151), where such needs are calculated through consideration of those factors which are accepted by the States concerned as relevant to water allocation. Therefore, the factors considered relevant to understanding each State's dependence on the shared waters, and thus to the calculation of each State's equitable and reasonable allocation of uses and benefits, are absolutely central, and codified or conventional formulations of the principle usually include an accompanying indicative list of such relevant factors. Such a list was first set out in Article V(2) of the 1966 Helsinki Rules. Most notably today, Article 6(1) of the UNWC now lists the following factors as relevant:

(a) Geographic, hydrographic, hydrological, climatic, ecological and other factors of a natural character;
(b) The social and economic needs of the watercourse States concerned;
(c) The population dependent on the watercourse in each State;
(d) The effects of the use or uses of the watercourses in one watercourse State on other watercourse States;
(e) Existing and potential uses of the watercourse;

(f) Conservation, protection, development and economy of use of the water resources of the watercourse and the costs of measures taken to that effect;
(g) The availability of alternatives, of comparative value, to a particular planned or existing use.

This list is not intended to be exhaustive and a range of additional factors might be relevant in the particular circumstances of a particular basin, negotiation or dispute, such as any religious, cultural or local customary significance attached to the river in question or to its waters. Similarly, the conduct of the States concerned regarding a contested use or project might be relevant including, for example, excessive delay in raising objections (McIntyre 2007, p. 186–189).

While all key instruments of international water law emphasise the lack of a hierarchy among the relevant factors (UNWC 1997, Article 6(3); ILA 1966, Article V(3)), it is apparent from the practice of States that certain considerations will usually be accorded more significance than others. For example, while Article 6(3) of the UNWC provides that '[t]he weight to be given to each factor is to be determined by its importance in comparison with that of other relevant factors', Article 10(2) would appear to prioritise 'vital human needs', a key element in identifying the 'population dependent on the watercourse in each State' as a relevant factor under Article 6(2). Of course, this elevation of vital human needs is 'likely to enhance the "human right dimension" of the use of the waters of international watercourses' (Tanzi and Arcari 2001, p. 131; Tully 2003, p. 101). A statement of understanding agreed by the UN General Assembly Working Group at the time of the adoption of the Convention advises that 'special attention is to be paid to providing sufficient water to sustain human life, including both drinking water and water required for production of food in order to prevent starvation', a position consistent with the ongoing discourse in international law on the human right to water (McIntyre 2015, p. 345).

Indeed, it would appear from the practice of States in this field that what matters above all else is the dependence of each watercourse State upon the shared waters in question, in terms of either human, social, economic or environmental needs, and that the relevant factors listed above and elsewhere largely function to elucidate the true nature and extent of such dependence (Fuentes 1996). For example, though the UNWC suggests that existing and potential uses of a watercourse will in principle be considered equally, with the ILC noting that 'neither is given priority' and that 'one or both factors may be relevant in a given case' (ILC 1994, p. 233), it is likely that existing uses will be favoured in practice as they can more easily be scrutinised in terms of their human, social, economic or environmental benefits (or adverse impacts), while the difficulties inherent in reliably considering the beneficial character (or negative impacts) of future uses are manifest (McIntyre 2007, p. 165; Lipper 1967, p. 50; Jiménez de Aréchaga 1960, pp. 335–335). Equally, the examination of factors such as efforts at conservation and economy of use of water resources by a particular State and the availability to a State of alternatives to a planned or existing use of shared waters primarily help to inform that State's true dependence upon the contested waters (McIntyre 2007, pp. 173–179). Further, though 'natural' factors, including the geography and hydrology of the basin are listed first under both the 1997 UNWC and the 1966 Helsinki Rules, there is general agreement among scholars that such factors are of only marginal significance as these do not relate directly to a State's dependence on the shared water and so could undermine the distributive nature of the equitable allocation envisaged under the principle of equitable and reasonable utilisation. For example, Tanzi and Arcari suggest that to accord any *a priori* pre-eminence to such circumstances 'would prejudice the principled equality among riparians' (Tanzi and Arcari 2001, p. 124). Similarly, Lipper argues convincingly that '[e]quality of right is the equal right of each co-riparian state to a division of the waters on the basis of its economic and social needs, consistent with the corresponding rights of its co-riparian states, and excluding from consideration factors unrelated to such needs' (Lipper 1967, p. 63). Thus, equality of right does not entitle each State to an equal share in the waters of a shared basin, but only to an equal right *vis-à-vis* its co-riparian neighbours to an equitable share of the uses and benefits of the watercourse having regard to all relevant factors. In essence, the principle of equitable and reasonable utilisation requires that, in using shared water resources, each co-basin State must have equitable and reasonable regard for the legitimate needs and interests of other co-basin States. The distributive nature of equity as applied in the particular field of international water law is highlighted by the fact that the significance attributed to the physical characteristics of the drainage basin, such as the length of the course of a river situated within each basin State, the extent of the drainage basin area lying in the territories of the basin States, or their relative contribution of water to the flow of a river, is relatively low (Fuentes 1996, p. 395–408). This situation contrasts with the application of equitable principles in maritime territorial delimitation, where the emphasis has been placed on the extent of each State's coastline (McIntyre 2007, p. 137–142).

However, such "rules of thumb" regarding the relative value accorded to different relevant factors tell us little about the nature of equity as a source of rules impacting on the determination of watercourse States' right to utilise shared waters, or about the means by which equitable principles can be operationalised in the practice of international water law. As this author has elsewhere suggested that 'one reason for confusion about the precise meaning of "equity" in the area of international water law is that it has historically been invoked in relation to a number of different roles' (McIntyre 2013a, p. 116), it is helpful to try to identify and to differentiate between several of the roles that equity might play as a body of normativity in respect of the use and protection of shared international water resources.

3.3 Functions of Equity in International Water Law

A wide range of functions can be attributed to equity in international law generally (Lowe 1992; Lapidoth 1987) and even in the narrow field of international water law, multiple overlapping and interrelated roles can be identified (McIntyre 2013a). For example, citing Thomas Franck (1995), scholars have recently characterised "fairness", which they describe as 'closely related, if not synonymous [to] principles of equity and justice', within an international watercourse context, as 'comprising two key elements: legitimacy and distributive justice' (Yihdego and Rieu-Clarke 2016, p. 529). However, this chapter focuses on a few of the principal ways in which equity can operate, within the rubric of international law, to facilitate the fair sharing of benefits deriving from shared water resources and to ensure effective environmental protection of international watercourses and their dependent ecosystems. Thus, it will not explore, for example, the notional competence of the International Court of Justice (ICJ) under Article 38(2) of its Statute to apply, with the consent of the parties to a dispute, equity *ex aequo et bono*, which commentators understand as not referring to rules of law, primary or supplemental, but to the Court's capacity to settle disputes on the basis of conciliation (Goldie 1987, p. 107; Lapidoth 1987, p. 172; Lowe 1992, p. 56). In applying equity *ex aequo et bono* 'the Court would have to decide according to non-legal principles of justice, of morality, of usefulness, of political prudence, and of common sense' (Berber 1959, p. 266–267), and so this mode of dispute settlement does not refer to considerations lying within the rules of law and, as such, does not form a component part of the corpus of rules and principles that constitute international law (McIntyre 2013a, b, p. 117). As parties to disputes are understandably reluctant to give the Court such wide and unfettered discretion, neither the ICJ nor its predecessor, the Permanent Court of International Justice (PCIJ), has ever decided a case *ex aequo et bono*. In the 1929 *Free Zones Case*, the PCIJ was careful to disregard any consideration of equity *ex aequo et bono* in the absence of the clear and explicit agreement of the parties.[4] In employing the concept of equity, the ICJ has made a point of clarifying that it is referring to a role other than that of equity *ex aequo et bono*:

> Whatever the legal reasoning of a court of justice, its decision must by definition be just, and <u>therefore</u> in that sense equitable. Nevertheless, when mention is made of a court dispensing justice or declaring the law what is meant is that the decision finds its objective justification in considerations lying not outside but within the rules ... There is consequently no question in this case of any decision *ex aequo et bono*.[5]

Exploring the principal ways in which equity can operate in the field of international water law involves characterising equity in a number of legal roles, including as a general principle of international law, as a means of ensuring the procedural fairness of inter-State communication and engagement arrangements, and as a substantive rule of apportionment of water quantum or uses. Of course, these roles are not mutually exclusive, but merely reflect the primary way in which equitable principles tend to be invoked and applied in respect of international freshwater resources.

---

[4] *Free Zones Case (France v. Switzerland)*, (1929) PCIJ Series A, no. 22, p. 5.

[5] *North Sea Continental Shelf Cases (Germany/Denmark/Netherlands)*, (1969) *ICJ Reports* 50, Judgment of 20 February 1969, at 3 and 47.

3.3.1 Equity as a General Principle of Law

The place of equity among the rules of international law is commonly understood as that of a general principle of law. The drafters of the ICJ Statute considered the 'general principles of law recognized by civilized nations' as belonging among the sources of international law 'in virtue of their social foundation and rational character', while Article 38(1)(c) of the Statute accommodates the evolution of general legal principles as they are formed in national legal systems through the ongoing clarification of the central idea of justice and the implementation of this idea into rules (Goldie 1987, pp. 105–106). In the *River Meuse* case, Judge Anzilotti said of the principle inherent in the Roman law maxim *inadimplenti non est adimplendum* (he who fails to fulfil his part of an agreement cannot enforce that bargain against the other party) that it is 'so just, so equitable, so universally recognised, that it must be applied in international relations also'.[6] In the same case, Judge Hudson observed that under 'Article 38 of the Statute, if not independently of that Article, the Court has some freedom to consider principles of equity as part of the international law which it must apply'.[7]

As regards the reference to 'general principles of law' in Article 38(1)(c), uncertainty persists as to whether it is intended solely to permit the Court to apply widely employed principles of national law where there might otherwise be lacunae among the established rules of international law, or also to include the various 'principles' of international law commonly included in natural resources and environmental treaties and declarative instruments, most notably the 1972 Stockholm Declaration and the 1992 Rio Declaration. While the latter category would include those guiding principles of environmental law routinely endorsed by States in their conventional and declarative practice, such as the precautionary principle, the polluter-pays principle, and the principle of common but differentiated responsibility, the former would include general principles of "natural justice" 'accepted by all nations in *foro domestico*', which could operate 'to avoid any possibility of a *non liquet* where there may be gaps in the law' (Birnie et al. 2009, pp. 26–27). The doctrines of abuse of rights and good faith are often cited as examples of such 'general principles'.[8] However, the most prominent, and normatively rich, of such general principles of natural justice is that of "equity", which plays a particularly significant role in the establishment, operation and application of the rules of international natural resources law (McIntyre 2013a, p. 112; Franck 1995, p. 56), and may be defined in this context as 'considerations of fairness, reasonableness, and policy often necessary for the sensible application of the more settled rules of law' (Brownlie 1990, p. 26; Lowe 1992, p. 54). As the concept of equity and particular equitable principles

[6] *Diversion of Water from the River Meuse (Netherlands v. Belgium)*, (1937) PCIJ Series A/B, No 77, at 50.

[7] *Ibid.*, at 77.

[8] *Case of the Free Zones of Upper Saxony and the District of Gex (Switzerland v. France)*, (1929) PCIJ Rep Series A/B No 46, at 167.

are to be found in many national legal systems, equity can undoubtedly play a role as a component of the corpus of norms that constitute international law (Lowe 1992, p. 55). That international tribunals may be entitled to apply equitable principles without the express authorisation of the parties to an inter-State dispute was confirmed by Judge Hudson in the *River Meuse* case, where he stated that '[w]hat are widely known as principles of equity have long been considered to constitute a part of international law, and as such they have often been applied by international tribunals'.[9]

However, leading commentators caution that, rather than borrow mechanically from domestic law, tribunals have only 'invoked elements of legal reasoning and private law analogies', so that 'general principles derived by analogy from domestic law are only marginally useful in an environmental context' (Birnie et al. 2009, p. 27). Agreeing that their role has been marginal in the development of international natural resources and environmental law, Maljean-Dubois suggests that, 'with the purpose of filling the gaps in conventional or customary law, these principles [should] play, *a priori*, a more important role in new fields such as environmental protection than in more traditional fields' (Maljean-Dubois 2011, p. 44). Somewhat ironically, the marginal utility of traditional domestic equitable principles is illustrated inadvertently by Judge Hudson who, in his separate opinion in the *River Meuse* case, cites several of the traditional maxims of equity found in Anglo-American jurisprudence, which he regarded as potentially relevant to transboundary water resource disputes, including "he who comes to equity must come with clean hands", "he who seeks equity must do equity", and "equality is equity".[10] The first of these maxims might apply to require that a party to a dispute seeking a remedy under international law ought to have acted in good faith and have discharged all relevant procedural and substantive obligations. In turn, the second might mean that the State that exploits the shared resource first may not object when the neighbouring State begins to do so or, conversely, that the state that succeeds in preventing the exploitation of the shared resource by a neighbour may itself be estopped from exploiting the resource (Botchway 2003, p. 217). This is essentially what occurred in the *River Meuse* case, where the Netherlands' complaint against Belgium's diversion of their shared River Meuse was dismissed largely because the Netherlands itself had earlier engaged in a similar diversion scheme. The third equitable maxim cited by Judge Hudson might suggest that there ought to be a proportionate distribution of benefits and burdens in the use of shared resources and might, in practical terms, ensure that 'equality would promote equity by the reliance on objective criteria that correspond with need, capacity, and symbiotic mutuality' (Botchway 2003, p. 217). Other maxims which may be of relevance to the equitable and reasonable utilisation of shared resources include "equity will not suffer a wrong to be without a remedy", which might influence the application of rules on state responsibility and liability. Similarly, the maxim "equity imputes an obligation to fulfil an obligation" might influence the application of rules on the enforcement of conventional obligations. However, Lowe cautions generally

[9] *Diversion of Water from the River Meuse (Netherlands v. Belgium)* PCIJ Series A/B No 70, 76–77.
[10] *Ibid.*, at 77.

about 'the difficulty of drawing equitable principles from national legal systems and applying them in the international system' (Lowe 1992, p. 80), and it is immediately apparent that not all equitable maxims will apply in the context of international watercourse law. Accordingly, one should be wary of assuming their relevance. For example, the maxims "where the equities are equal, the first in time shall prevail" and "delay defeats equity" might appear to support the doctrine of prior appropriation which, in the case of international watercourses, has largely been contradicted by Article 6 of the UNWC and anyway had not been supported in state practice (Fuentes 1996, p. 365). It has also been widely criticized as wasteful, not conducive to the optimal economic development of the watercourse, and potentially environmentally damaging (Lipper 1967, p. 51). Similarly, the maxim "where there is equal equity, the law shall prevail" might incorrectly be assumed to suggest that the established status quo in transboundary resource regimes should not be disturbed (Botchway 2003, p. 218).

As a general aspirational ideal, the notion of equity is also of limited utility. Although supportive statements from judicial and other actors characterize equity as a 'direct emanation of the idea of justice' or as necessary 'considerations of fairness, reasonableness and policy', they provide only the vaguest guidance for legal decision-makers regarding relevant social values. According to the ICJ,

> It is not a question of applying equity simply as a matter of abstract justice, but of applying a rule of law which itself requires the application of equitable principles, in accordance with the ideas that have always underlain the development of the legal regime.[11]

Such statements do little to help us to understand how equitable principles might be applied by international lawyers in the settlement of international disputes over natural resources. However, international courts have employed equitable principles to resolve natural resource disputes on a number of occasions, particularly in respect of maritime boundaries and resources (Charney 1984; Higgins 1991; Nelson 1990). Notable examples include the 1969 *North Sea Continental Shelf* cases,[12] where the ICJ resorted to the formulation of equitable principles concerning the lateral delimitation of adjacent areas of the continental shelf in the absence of customary or treaty law rules which bound the State parties to the dispute; the 1974 *Fisheries Jurisdiction Case (United Kingdom v. Iceland)*,[13] where the ICJ outlined the elements of an "equitable solution" to a dispute over fishing rights and directed the parties to negotiate accordingly; the 1975 *Anglo-French Continental Shelf Arbitration (UK v. France)*,[14] concerned with the territorial division of the English Channel; the 1982 *Tunisia-Libya Continental Shelf* case[15]; the 1984 *Gulf of Maine* case,[16] which concerned the

[11] *North Sea Continental Shelf Cases (Germany/Denmark/Netherlands)*, (1969) *ICJ Reports* 50, Judgment of 20 February 1969, p. 47, para. 85.

[12] *North Sea Continental Shelf Cases (Germany/Denmark/Netherlands)*, (1969) *ICJ Reports* 50.

[13] (1974) *ICJ Reports* 3.

[14] 54 *ILR* 6 (1975); (1979) 18 *ILM* 397.

[15] (1982) *ICJ Reports* 18.

[16] (1984) *ICJ Reports* 246.

delimitation of fisheries zones and the subsoil of the continental shelf; the 1985 *Libya-Malta Continental Shelf* case[17]; the 1985 *Guinea–Guinea-Bissau Arbitration*[18]; the *Case Concerning the Frontier Dispute* (*Burkino Faso/Mali*) (1986),[19] where the ICJ used equity to decide on the division of a frontier pool; and the 1993 *Case Concerning Maritime Delimitation in the Area between Greenland and Jan Mayen*.[20]

However, the cases listed above have not elaborated greatly on how equity applies as a general principle of law. Rather, the ICJ has taken greater care to clarify precisely what roles equity as a general principle of law cannot play. For example, the Court has distinguished between the application of equity in international law and the use of the term in the context of some domestic legal systems, where equity serves to ameliorate the overly rigorous and harsh application of rules of law in order to ensure justice. In this role it can be contrasted with rigid rules of law. In the *Tunisia-Libya Continental Shelf* case, the ICJ stated that 'in general, this contrast has no parallel in the development of international law' and that 'the legal concept of equity is a general principle directly applicable *as law*'.[21] Once again the court emphasized that international equity must lie within the rules of law. Of course, the ICJ has been particularly careful to distinguish between the application of equity as part of the general principles of law and its application *ex aequo et bono* and, in so doing, it has emphasized that equitable considerations in the former sense lie within the rules of law.[22] However, not all commentators are convinced that the distinction between equity as a general principle of law and equity *ex aequo et bono* can be easily maintained. For example, Brownlie, though content with Judge Hudson's application of the principles of equity in the *River Meuse* case, is highly critical of later applications (Brownlie 1979, pp. 287–288). He suggests that the equitable principles laid down by the ICJ in the *North Sea Continental Shelf* cases and further developed by the Court of Arbitration in the 1978 *Western Approaches Arbitration*[23] 'amount to no more than a bundle of highly impressionistic ideas' and, further, that when 'employed in this way "equitable principles" become highly faint indications of the reasoning … on which judicial discretion has been exercised and may be exercised in other cases' (Brownlie 1979, p. 287). He concludes more generally that, whatever 'the particular and interstitial significance of equity in the law of nations, as a general reservoir of ideas and solutions for sophisticated problems it offers little but disappointment' (Brownlie 1979, p. 288).

Equity as a general principle of law may play an important role in facilitating the rational and structured integration and reconciliation of different objectives or values which have traditionally been understood to be in competition with each other.

---

[17] (1985) *ICJ Reports* 13.

[18] (1985) XIX *RIAA* 149.

[19] (1986) *ICJ Reports* 554.

[20] (1993) *ICJ Reports* 38.

[21] *Supra*, n. 15, at 60, para. 71.

[22] *North Sea Continental Shelf Cases, supra*, n. 5, at 47, para. 85.

[23] HMSO Misc. No. 15 (1978), Cmnd. 7438.

Equity is, after all, commonly understood as a means of taking account of all relevant circumstances in a particular case (Higgins 1994, p. 221), while the concept of sustainable development has its origins in conventional and declaratory instruments of international environmental law as a means of reconciling protection of the natural environment with the requirements of economic and social development. In relation to the use of shared freshwater resources, it is widely accepted that the principle of equitable and reasonable utilisation "operationalises" the concept of sustainable development (Kroes 1997, p. 83; McIntyre 2007, p. 247; Wouters and Rieu-Clarke 2001, p. 283). In analysing the principles of equity, no-harm and sustainability as included in the UNWC, Charles Bourne concludes that sustainability is a goal or objective which could only be attained by reliance on equity (Bourne 1997, pp. 221–230). Lowe concludes, in relation to the flexibility inherent in the application of equitable principles, that:

> These characteristics make equity particularly suitable for discussions in contexts where there are competing interests which have not hardened into specific rights and duties. This will be true primarily in areas where the law is not highly developed. The nascent concept of intergenerational equity, and of equitable principles in environmental law, are examples (Lowe 1992, p. 73).

Thus, equitable principles can be understood as a legal means of facilitating the integration of diverse values and objectives, including environmental values, into implementation of the multi-faceted principle of equitable and reasonable utilisation. This role involves the non-controversial application of equity *infra legem*, 'which constitutes a method of interpretation of the law in force, and is one of its attributes',[24] and has been characterised as equity 'used to adapt the law to the facts of individual cases' (Akehurst 1976, p. 801). The ICJ appears to have applied equity in this manner in the *Tunisia-Libya Continental Shelf* case, observing that 'when applying positive international law, a court may choose among several possible interpretations of the law the one that appears, in the light of the circumstances of the case, to be closest to the requirements of justice'.[25] However, Higgins expresses concern that employing any inherently subjective notion of the "requirements of justice" to inform the interpretation of legal rules 'is merely to avoid justifying and making specific certain policy objectives' (Higgins 1994, p. 220). Nevertheless, this role for equity is likely to become ever more apparent, and ever more important, as international water law struggles, not alone to take account of the myriad factors potentially relevant to equitable and reasonable utilisation of the kind indicated in Article 6(2) of the UNWC, but also to take on board the burgeoning normative implications of the increasingly pervasive requirements of ecosystems protection under international law (McIntyre 2004, 2014) or of the emerging human rights of access to adequate water and sanitation (Tully 2003; McIntyre 2015).

[24] *Frontier Dispute Case (Burkino Faso v. Mali)* (1986) *ICJ Reports* 554.

[25] *Supra*, n. 15, at 71.

3.3.2 Procedural Equity

It is increasingly apparent that the procedural rules of international water law play a particularly significant role in relation to the equitable and reasonable utilisation of shared waters, because they facilitate the orderly collection and communication of information according to agreed methodologies, which is absolutely vital to the equitable consideration of the interests of States. It is notable that in the 2010 *Pulp Mills* case,[26] the ICJ recognized the central role of procedural rules in the effective implementation of the substantive rules of international water law and, more particularly, that the environmental impact assessment process plays a key role in ensuring that social and environmental considerations relating to a planned or continuing use of an international watercourse are adequately understood and communicated so that they may properly be taken into account as a factor within the balancing process that lies at the heart of equitable and reasonable utilisation (McIntyre 2010, 2011, 2013b). Therefore, international water law tends to stress the equitable participation of all riparian States. For example, Article 4 of the UNWC seeks to ensure the right to participate of any watercourse State that might be significantly affected by a proposed watercourse agreement between other co-riparians, while Article 5(2) sets out the principle of "equitable and reasonable participation". Characteristically, while Articles 11–19 set out detailed rules on notification, consultation and negotiation in respect of "planned measures", Article 13(b) allows for extension of the period for reply to notification 'at the request of a notified State for which the evaluation of the planned measures poses special difficulty'. This provision typifies the imperative of equitably ensuring the meaningful and effective participation of all watercourse States in the process of international water law, which can be regarded as a practical application of the equitable maxim "equality is equity" (McIntyre 2013a, pp. 116–117).

It has long been understood that a normative framework requiring the equitable balancing of the legitimate interests of basin States must inevitably involve intense procedural inter-State engagement (Bruhacs 1993, p. 159), which often can only be facilitated by the establishment of technically competent inter-State institutional machinery. Such institutions can ensure effective inter-State communication which might involve, *inter alia*, prior notification of planned projects potentially impacting upon the watercourse, routine exchange of information regarding the utilisation or condition of the shared waters, or expression of concerns on the part of any basin State. The pivotal role of institutional mechanisms in giving effect to the principle of equitable and reasonable utilisation has long been recognised by the international community, with Recommendation 51 of the *Action Plan for the Human Environment* adopted at the 1972 Stockholm Conference calling for the 'creation of river basin commissions or other appropriate machinery for cooperation between interested States for water resources common to more than one jurisdiction', and setting down a number of basic principles by which the establishment of such bodies should be

[26] *Case Concerning Pulp Mills on the River Uruguay* (*Argentina v. Uruguay*), (2010) 152 *ILR* 1.

guided (UNCHE 1972). Although such institutional structures can take numerous different forms and have diverse remits, there are today at least 119 river basin organisations (RBOs) performing a very extensive range of coordination and joint management functions (Schmeier 2013, p. 65). Reliance on such institutional mechanisms to facilitate the inter-State cooperation necessary to achieve equitable and reasonable utilisation is often referred to as the "common management" approach, which further underlines the existence of a community of interest among co-basin States (Birnie and Boyle 1992, pp. 223–224; McIntyre 2007, pp. 28–40). Consistent with the principle of "equitable and reasonable participation" articulated under Article 5(2) of the UNWC, the rules of international water law ought to be interpreted and applied in such a manner as to require that basin States take all reasonable measures to facilitate the meaningful and effective participation of other basin States in such cooperative institutional mechanisms. In some circumstances, this might require, for example, the provision of financial or technical assistance in order to ensure a less capacitated State's equitable participation.

3.3.3 Equity as a Substantive Rule of Apportionment

When viewed as a stand-alone, substantive rule for the apportionment amongst States of the uses or benefits of shared international natural resources, three possible roles have been identified, which may be complementary and may occur concurrently: *i.e.* equity as a means of achieving a desired equitable result; equity as a process for taking account of all the relevant circumstances in a particular case; and equity as a means of rendering specific and applicable to a particular case laws of general application (Higgins 1994, pp. 220–222).

One of the roles of equity most widely employed by international courts and tribunals is that of choosing among possible interpretations of the law in such a way as to reach a just or equitable solution. The ICJ took this approach in the *Tunisia-Libya Continental Shelf* case,[27] where the court focused on achieving what it regarded as an equitable result. However, the court insisted that the search for an equitable result was not an operation of distributive justice but merely an operation of equity in a corrective role. This corrective function can only take place in a manner consistent with the rules of law and would never be acceptable *contra legem*. In the *Libya-Malta* case,[28] the ICJ again reiterated the distinction between this role of equity and the operation of distributive justice, where it listed as an example of an equitable principle the principle that there can be no question of distributive justice. However, while this argument might be tenable in maritime boundary delimitation, it would be a great deal more difficult to argue that the act of interpreting the rules of international water law so as to achieve an equitable result was not an operation of distributive justice. The cardinal and overarching principle of equitable and

[27] *Supra*, n. 15.

[28] *Supra*, n. 17.

reasonable utilisation has long been understood as a process for the allocation of rights in the uses and benefits of shared water resources on the basis of a distributive conception of equity having regard to all relevant factors, whereby uses and benefits will be shared in proportion to each basin State's needs, where such needs are calculated through consideration of factors such as those listed under Article 6(2) of the UNWC (McIntyre 2007, pp. 147–151).

A second conception of equity identified by commentators (Higgins 1994, p. 221), is that equity in international law lacks specific content but operates rather as a means for considering all the relevant circumstances in a particular case. In the *Tunisia-Libya Continental Shelf* case,[29] the ICJ seems to have supported this view, finding that it was 'virtually impossible to achieve an equitable solution to any delimitation without taking into account the particular relevant circumstances of the area'. In this context, it would appear that 'there are few, if any, constraints upon the factors which may form the basis of an argument in equity' (Lowe 1992, pp. 72–73). Lowe also points out that, according to the ICJ in *North Sea Continental Shelf*, 'there is no legal limit to the considerations which States may take account of for the purpose of making sure they apply equitable procedures'.[30] Emphasising the potential flexibility of equity in this role, he further observes that 'once the relevant factors have been considered the person making the decision is freed from the necessity of making the reasoning consistent with established legal rules and principles', though he does concede that 'even equity must be consistent'.[31] Equity in this role resonates with the pre-eminent international water law principle of equitable and reasonable utilisation which, as formulated under the Helsinki Rules (ILA 1966) and the UNWC, provides a non-exhaustive, indicative checklist of factors which are to be considered. However, neither formulation offers any guidance as to the weight or priority to be given to the various factors listed as relevant to equitable and reasonable utilisation, so that the principle gives little normative guidance as to what should happen in any particular situation. Instead, each provides, rather unhelpfully, that all the factors must be balanced with other factors and a decision made on the basis of the whole. Due to its normative vagueness, some writers have tended to be pessimistic about the principle's usefulness, though others feel it retains merit as a procedural approach (McCaffrey 2001, p. 345; Tanzi and Arcari 2001, p. 109).

The third role for equity of relevance to the application of the principle of equitable utilisation is that of establishing the specific content of rules which are too general or vague to be applied directly in certain circumstances. In this way, equity permits the application of general legal rules to specific, concrete situations. De Visscher envisages equity in this role, suggesting that "l'équité est la norme du cas individual" (Higgins 1994, p. 222). Therefore, equity might be expected to play a crucial role in elaborating the substantive content of the principle of equitable and reasonable utilisation. Looking more particularly at the application of equitable principles to shared natural resources, Franck recognises three distinct approaches

[29] *Supra*, n. 15.

[30] *Supra*, n. 5, at 72–73.

[31] *Ibid.*

to equitable allocation of shared resources: *i.e.* "corrective equity"; "broadly conceived equity"; and "common heritage equity" (Franck 1995, p. 57). Under the "corrective equity" approach, equitable considerations are only exceptionally invoked and function to ameliorate the gross unfairness which might occasionally result from the strict application of technical legal rules. This is the most conservative approach, confining the exceptional application of equitable principles within a dominant rule of resource allocation. Under the "broadly conceived equity" approach, equity itself comprises a rule of law and is the dominant applicable rule for resource allocation. This approach affords tribunals a great deal more discretion than corrective equity and tends to be more openly distributive. Franck regards the principle of equitable and reasonable utilisation, as incorporated into the UNWC, as an example of broadly conceived equity and as indicative of a recent trend to include similar equitable mechanisms in natural resource and environmental treaty regimes. Generally, Franck identifies a trend (exemplified by the adoption of Article 83(1) of the 1982 UN Convention on the Law of the Sea requiring states 'to achieve an equitable solution' in continental shelf delimitation) towards the introduction of broadly conceived equity into conventional provisions relating to the allocation of shared natural resources, which will increasingly compel courts and tribunals to apply broader notions of distributive justice (Franck 1995, pp. 61–75). "Common heritage equity" applies to the allocation of resources which are the patrimony of all humanity, such as outer space, Antarctica or the mineral resources of the deep seabed, and often involves a "trust" model in which conservation is the first or sole priority. Clearly this last approach has limited relevance for shared freshwater resources, where utilisation rights are associated with territorial sovereignty over a portion of the watercourse or basin.

From the above brief examination of the potential roles of equity in the application of the equitable and reasonable utilisation principle it becomes apparent that equity can, and usually will, play a combination of at least two, if not three, of the roles identified. Thus, it appears axiomatic that equity is applied at several stages, *i.e.* in the identification of a just and equitable solution, in the consideration of all relevant factors and circumstances, and in the concrete elaboration of normatively vague rules. However, no formulation of the principle offers guidance as to the order in which equity might perform these functions. For example, would a court identify a just result and go on to achieve this result by giving appropriately weighted consideration to each relevant factor, or would a just result be determined by prior consideration of the relevant factors? The former approach would afford a court a greater degree of discretion in establishing what constitutes an equitable result, and this result would necessarily be subjective. Similarly, where equity functions to elaborate specific rules for a particular case, it is unclear whether these rules determine the priority to be given to each relevant factor or whether prior consideration of the factors is necessary in order to determine the rules to be applied. What is clear is that equitable utilisation can only effectively function as a procedure, a fact long recognised by specialist commentators (Bruhacs 1993, p. 159).

An examination of the practice of international tribunals on maritime delimitation suggests, controversially, that the Court now begins by identifying what it considers to be an equitable result (McIntyre 2013a, pp. 122–124). In the *North Sea Continental Shelf* cases,[32] the Court decided in the circumstances only to consider a particular conception of geographical proportionality which stressed the respective length of each State's coastline, on the basis of which it adjusted the delimitation to account for the unfavourable concavity of the (then) West German coastline. At all times the Court presented its deliberations as entirely normative in character, suggesting that it had begun with and concentrated on equitable principles (McIntyre 2013a, p. 122). Similarly, in the *Anglo-French Continental Shelf* arbitration,[33] the arbitral tribunal considered a number of equitable factors in order to 'balance the equities', including defence considerations as well as the size of the population and political and economic importance of the Channel Islands, but primarily employed the equitable principle of proportionality to arrive at an equitable delimitation. Once again, the tribunal took care to emphasise that it was applying normative principles of justice rather than a more discretionary concept of distributive justice (Franck 1995, p. 64). The ICJ emphatically affirmed this approach in the *Fisheries Jurisdiction* case, stating '[i]t is not a matter of finding simply an equitable solution, but an equitable solution derived from the applicable law',[34] and thereby unequivocally rejecting the equitable solution as the starting point.

However, the court subsequently adopted the very opposite approach in the *Tunisia-Libya Continental Shelf* case, holding that it was 'bound to decide the case on the basis of equitable principles' but, in giving an account of what equitable principles entail, the court stated that:

> The result of the application of general principles must be equitable… It is not every such principle which is in itself equitable; it may acquire this quality by reference to the equitableness of the solution… "Equitable principles"… refers back to the principles and rules which may be appropriate in order to achieve an equitable result.[35]

This position reflects the position under Article 83(1) of the then newly adopted UN Convention on the Law of the Sea (UNCLOS), itself opened for signature in 1982, providing that 'delimitation of the continental shelf … shall be effected by agreement on the basis of international law … *in order to achieve an equitable solution*' (emphasis added). The Court found that the omission of any reference to technical rules of delimitation, such as "equidistance", meant that there was no longer any formal textual guidance as to the content of an equitable solution and, therefore that the goal of reaching an equitable result must determine the means for achieving it (Franck 1995, p. 86). The court stated that 'the equitableness of a principle must be assessed in the light of its usefulness for the purpose of arriving at an equitable result'.[36] The Court exercised considerable discretion and took into account a wide range of factors, including the general configuration of the coastlines, the existence

---

[32] *Supra*, n. 5.

[33] *Supra*, n. 14.

[34] *Supra*, n. 13, at 33.

[35] *Supra*, n. 15, at 70.

[36] *Tunisia-Libya Continental Shelf* case, *ibid.*, at 59.

and position of various islands, the configuration of the land frontier and the conduct of the parties in the granting of petroleum concessions. Primarily, however, the Court again took account of proportionality having regard to the respective length of the parties' coastlines. In the *Gulf of Maine* case, which involved delimitation of fisheries zones and the subsoil of the continental shelf, the Court once again emphasized the requirement of an equitable solution and adopted proportionality as 'the primary tool for the application of broadly conceived equity' (Franck 1995, p. 69). It held that 'delimitation should be effected by the application of equitable criteria capable of ensuring … an equitable result'.[37] This "broadly conceived equity" approach was followed in the *Libya-Malta Continental Shelf* case, where the Court affirmed the primacy of the equitable result over those equitable principles used to achieve it. It referred to the equitable result as 'the primary element in this duality of characterization'[38] and stressed the role of proportionality in achieving such a result, describing it as 'intimately related… to the governing principle of equity'.[39]

The more recent approach of international courts and tribunals to the application of equity in maritime delimitation disputes – that of focusing on the desired result – has been almost universally criticized by commentators, most of whom fear it allows too great a degree of judicial discretion. Higgins, for example, while conceding that 'decisions will in reality, and necessarily reflect policy preferences', takes the view that 'these policy preferences should be articulated and tested against stated desired outcomes. In this way the objectives would be transparent and the methods objectively verifiable' (Higgins 1994, p. 224). Instead, she feels that the result-oriented approach taken by the court 'has allowed the Court to insist it is applying "an actual rule of law" – but one that is opaque and not capable of scrutiny or review' (Higgins 1994, p. 224). Jennings has also expressed concern, and almost goes so far as to question whether such an application of equity can be distinguished from equity *ex aequo et bono*, asking, '[i]s equity then just the lawyers' name for subjective judicial decision' (Jennings 1986, p. 31). Judge Gros, in his dissenting opinion in the *Gulf of Maine* case,[40] expressed his concern that equity must be controlled if it is to be predictable. Therefore, it is important to examine which criteria the court has used and to what extent each has been relied upon in deciding what constitutes an equitable result.

In the *North Sea Continental Shelf* cases,[41] the Court's conclusion that equidistance was inequitable appears to have been based principally on consideration of the natural or physical characteristics of each State's coastline. In particular, the court placed great importance on the relative lengths of States' coastlines. Though critics have questioned the validity of coastline length as a determining factor in continental shelf delimitation (Friedmann 1971, p. 757; Higgins 1994, p. 226; Rothpfeffer

[37] *Supra*, n. 16, at 300.

[38] *Supra*, n. 17, at 29.

[39] *Ibid.*, at 43.

[40] *Supra*, n. 16.

[41] *Supra*, n. 5.

1972, p. 115), in the *Libya-Malta* case[42] the ICJ listed among the equitable principles to be considered the principle that there is no question of refashioning geography or seeking to make equal what is unequal. This appears an unambiguous rejection of the possibility of the equitable result doctrine operating as a form of distributive justice, at least in the field of maritime delimitation, especially as the Court expressly refused to attach any legal significance to the respective economic needs of the parties, reasoning that delimitation should not be based on as transient a factor as the relative prosperity or poverty of the States at a particular point in time (McIntyre 2013a, p. 123). Similarly, in the *Gulf of Maine*[43] case the Court refused to attribute any significance to economic factors advanced by Canada, such as the special dependence of certain Nova Scotia communities on fishing, though it did assign a subordinate, corrective role to such factors, which could provide a *post hoc* check on the equitableness of a result achieved on the basis of geophysical proportionality. In the *Libya-Malta Continental Shelf* case,[44] while the Court considered itself free to take account of a range of factors, it once again refused to attach significance to economic needs, and so rejected Malta's claim that its lack of energy resources, its requirements as an island developing State, and the range of its fisheries should be considered. However, in recent years it appears that the court is increasingly likely to consider economic factors to some extent in determining an equitable maritime delimitation. In the *Jan Mayen* case,[45] the court considered the commercial value of the relevant fisheries in deciding that an equitable result required a larger area of the common shelf and fisheries zone, not strictly proportionate to the States' respective coastlines, to be allocated to Greenland (Franck 1995, p. 73). Of course, in the field of international water law, social and economic dependence on the contested water resources in question is likely to prove to be the key factor in determining equitable and reasonable utilisation, demonstrating the inherently distributive nature of equity in this field, which simply reflects peoples' unique and total social and economic dependence upon water as a natural resource.

3.4 Equity as Proportionality

In the area of territorial delimitation and shared resource allocation, proportionality has been described as 'one technique among many to achieve an equitable outcome in the face of special geographic circumstances' (Higgins 1994, p. 230). For the purpose of this chapter it is necessary first to trace the development of the concept through various continental shelf and maritime delimitation disputes before speculating on its potential role in determining what constitutes equitable and reasonable utilisation of shared water resources.

---

[42] *Supra*, n. 17.

[43] *Supra*, n. 16.

[44] *Supra*, n. 17.

[45] *Supra*, n. 20.

3.4.1 Proportionality and Maritime Delimitation

The concept of proportionality was employed by the ICJ in the *North Sea Continental Shelf* cases,[46] where it was regarded as an element of equity. In this instance the States concerned were called upon to recognise a reasonable degree of proportionality to determine 'the extent of the continental shelf appertaining to the States concerned and the lengths of their respective coastlines' (Goldie 1987, pp. 118–119). Rejecting the use of equity to promote distributive justice, the ICJ held that the role of equity, and of proportionality as an element of equity, was to correct anomalies rather than to ensure fair or equal shares. Therefore, the Court did not see its function as the apportionment of a just and equitable share of the divisible area, but merely the demarcation of boundaries in the continental shelf, where equidistance is not to be used, on the basis that there should exist some relationship between the amount of shelf awarded and the relative length of coastlines. Equity, and the equitable principle of proportionality in particular, was useful in 'abating the effects of an incidental special feature from which an unjustifiable difference of treatment could result'.[47] The Court, therefore, did not suggest that proportionality should form a rule of general applicability under which any coastal State could claim a share of continental shelf proportionate to the length of its coast, but instead limited its application as an aid to delimitation to where a balance must be struck between States with 'markedly concave or convex' or 'very irregular' coastlines (McCrae 1981, p. 292). Generally, in accordance with Article 6(2) of the 1958 Geneva Convention, the Court would seek to adhere to the equidistance line except where special circumstances would radically distort the boundary. Also, it is clear that the Court did not view proportionality 'as a distinct principle of delimitation, but as one of the factors to be considered in ensuring that equitable procedures were applied' (Higgins 1994, p. 229). The tribunal in the *Anglo-French Continental Shelf Arbitration*[48] was even less enthusiastic and found that proportionality was 'not a general principle providing an independent source of rights to areas of the continental shelf'. It expressed the view that the concept of proportionality was relevant only in negative terms, saying that 'it is disproportion rather than any general principle of proportionality which is the relevant criterion or factor'.[49] Therefore, proportionality would not determine the delimitation but could act as a *post hoc* check on the equity of a proposed solution and correct it where anomalies or distortions occurred. According to Franck, 'the tribunal seemed to adopt a notion of distributive justice, yet to contain it within the then-legitimate rule of equidistance' (Franck 1995, p. 65).

[46] *Supra*, n. 5.

[47] *Ibid.*, at 50.

[48] *Supra*, n. 14.

[49] *Ibid.*, at 58.

The ICJ gave its strongest endorsement to the proportionality principle in the 1982 *Tunisia-Libya Continental Shelf* case,[50] which reflected the elevation of the achievement of an equitable result as the new key objective of relevant conventional law under Article 83(1) of the 1982 UNCLOS and saw the Court revisiting its earlier application of proportionality in continental shelf cases. It stated that the question is one of 'proportionality as a function of equity' and that 'the element of proportionality is related to the lengths of the coasts concerned'.[51] According to Higgins, the Court in this instance, 'is essentially using proportionality as a substantive principle of delimitation' rather than confining its application to a *post hoc* check on the equitableness of a result reached by other means (Higgins 1994, p. 230). Furthermore, the Court appeared to suggest that where delimitation is based on proportionality, the Court is entitled to take a very flexible approach, thus enhancing its general discretion (Goldie 1987, p. 121). In the *Gulf of Maine* case,[52] proportionality became in effect the primary legal principle for the identification of an equitable solution, and in the *Libya-Malta* case[53] it was once again the dominant consideration, despite the absence of any distorting natural feature. Indeed, the Court held that the principle enjoyed two roles: both as a factor to be considered in the initial delimitation and as a *post hoc* check on the equitableness of the solution proposed, however that proposal was reached. According to Franck, who concludes that proportionality is 'now evidently the preferred means by which to reify the abstract notion of equity',

> the former confers upon the court the discretion to allocate resources according to considerations of fairness, while the latter allows the court to ensure that the result, achieved by reference to a range of considerations, is not unfairly influenced by the effect given to any one of them (Franck 1995, p. 71).

3.4.2 Proportionality and Shared International Water Resources

Whilst the approach taken by international tribunals to the role of proportionality in maritime delimitation is in many regards quite narrow and restrictive, the concept is certain to play a more fundamental and far-reaching role with regard to shared water resources. The law on non-navigational uses of international watercourses differs from that on continental shelf delimitation in that thorough codification has actually occurred, thereby giving a solid legal basis to those considerations against which proportionality, as a function of equity, can be measured. It should also be remembered that similar formulations of the principle of equitable and reasonable utilisation, complete with indicative lists of relevant considerations, have received

[50] *Supra*, n. 15.

[51] *Ibid.*, at 76.

[52] *Supra*, n. 16.

[53] *Supra*, n. 17.

considerable and widespread support in State practice. One might assume from the practice of international tribunals outlined above that, in identifying an equitable regime for the utilisation of an international watercourse, tribunals would emphasise the importance of the natural and physical characteristics of the watercourse within each State. Principal among such factors would be the extent of the drainage area within the territory of each State and the quantum contribution of water by each State to the flow of the watercourse. Indeed, the indicative list of factors relevant to equitable and reasonable utilisation provided under Article 6(1) of the UNWC hints that this may be so, as the first set of factors listed includes 'geographic, hydrographic, hydrological, climatic, ecological and other factors of a natural character'. However, on the basis of a very comprehensive survey of State practice and the decisions of arbitral and judicial tribunals relating to the allocation of shared freshwaters, Fuentes persuasively concludes that the significance attributed to the physical characteristics of the drainage basin, in particular the extent of the drainage area lying within the territories of the parties and their contribution of water to the flow of the river, is relatively low (Fuentes 1996, pp. 394–408). She points out, for example, that the Narmada Tribunal made a *prima facie* equitable apportionment on the basis of the social and economic needs of the parties, which was then modified slightly after consideration of the physical characteristics of the drainage basin, an accommodation which 'shows that the role ascribed to that [physical] criterion is low in the hierarchy of relevant factors' (Fuentes 1996, p. 408). She goes on to assert that this practice of attributing low significance to the physical characteristics of a drainage basin as factors in establishing an equitable regime for the utilisation of shared water resources 'is consistent with the rule of equitable utilization, because to admit these factors to function as a direct basis for the allocation of waters would not be in keeping with the principle of equality between the basin States' (Fuentes 1996, p. 408; Tanzi and Arcari 2001, p. 124). At any rate, by attaching more significance to the water needs of watercourse States than to the physical characteristics of the drainage basin, equity as applied in international freshwater law is considerably more distributive in nature than equity as applied in the law of continental shelf and maritime territorial delimitation.

3.5 Conclusion

This chapter has attempted to demonstrate the obvious parallels between the principle of equitable and reasonable utilisation and 'the doctrine of "equitable principles – equitable result" consistently developed and applied by the ICJ in continental shelf delimitation cases' (Tanzi and Arcari 2001, p. 98). It becomes apparent that the jurisprudence developed by the ICJ on the principle of proportionality as a function of equity should help to ensure that a use of an international watercourse by one watercourse State, no matter how beneficial, would not be permitted where it would result in a grossly disproportionate adverse impact on the social, economic, developmental or environmental requirements of other watercourse States. There can be little doubt, therefore, that equity as a body of normativity has a crucially significant role to play in the law relating to the allocation of shared water resources. Franck lauds the tempering role' that equity can play 'when the apportionment of goods … occurs in the context of an almost infinite number of possible geographical, geological, topographical, economic, political, strategic, demographic, and scientific variables' (Franck 1995, p. 79). Similarly, Lowe highlights the inherent flexibility of equitable principles and their ability to allow a broadening of the scope of enquiry (Lowe 1992, p. 73). These features clearly demonstrate the suitability of equity as the cornerstone of modern international water resources law (Fuentes 1996, p. 411).

By enhancing the legitimacy of the rules and related outcomes of international water law, both its procedural legitimacy, in the form of rules on inter-State communication, and its substantive legitimacy, in the form of distributive justice, the concept of equity enhances the law's effectiveness and, thus, can function to maximise water security. Commentators have long agreed that the perceived fairness of a particular rule of international law, or of the outcome produced by that rule, will influence that rule's "compliance pull" (Franck 1988, p. 706; Yihdego and Rieu-Clarke 2016, p. 528–529), and thereby its role in ensuring human well-being by achieving the greatest attainable measure of water security.

References

Akehurst, M. (1976). Equity and general principles of law. *International and Comparative Law Quarterly, 25*, 801–825.

Berber, F. J. (1959). *Rivers in international law*. London: Stevens.

Birnie, P., & Boyle, A. E. (1992). *International law and the environment*. Oxford: Oxford University Press.

Birnie, P., Boyle, A., & Redgwell, C. (2009). *International law and the environment* (3rd ed.). Oxford: Oxford University Press.

Botchway, F. N. (2003). The context of transboundary energy resource exploitation: The environment, the state and the methods. *Colorado Journal of International Environmental Law and Policy, 14*, 191–240.

Bourne, C. B. (1997). The primacy of the principle of equitable utilization in the 1997 watercourses convention. *Canadian Yearbook of International Law, 35*, 215–232.

Brownlie, I. (1979). Legal status of natural resources in international law (Some aspects). *162(I) Recueil des Cours, 249*.

Brownlie, I. (1990). *Principles of public international law* (4th ed.). Oxford: Oxford University Press.

Bruhacs, J. (1993). *The law of non-navigational uses of international watercourses*. Dordrecht: Martinus Nijhoff.

CBD. (1992). *United Nations convention on biological diversity*. 1760 UNTS 79.

Charney, J. I. (1984). Ocean boundaries between Nations: A theory for progress. *American Journal of International Law, 78*, 582–606.

Committee on Economic, Social and Cultural Rights (CESCR). (2002, November 26). *General comment No. 15: The right to water: Articles 11 and 12 of the international covenant on economic, social and cultural rights*. U.N. DOC. E/C.12/2002/11.

De Aréchaga, E. J. (1978). International law in the past third of a century. *Recueil des Cours, 159*, 9.

Franck, T. M. (1988). Legitimacy in the international system. *American Journal of International Law, 82*, 705–759.
Franck, T. M. (1995). *Fairness in international law and institutions*. Oxford: Clarendon Press.
Friedmann, W. (1971). Selden Redivivus – Towards a partition of the seas? *American Journal of International Law, 65*, 757–770.
Fuentes, X. (1996). The criteria for the equitable utilization of international rivers. *British Yearbook of International Law, 65*, 337–412.
Goldie, L. F. E. (1987). Equity and the international management of transboundary resources. In A. Utton & L. Teclaff (Eds.), *Transboundary resources law*. London: Westview Press.
Higgins, R. (1991). International law and the avoidance, containment and resolution of disputes. *Recueil des Cours, 230*, 9–341.
Higgins, R. (1994). *Problems and process: International law and how we use it*. Oxford: Clarendon Press.
International Law Association (ILA). (1966). Helsinki rules on the uses of the waters of international rivers. In *Report of the fifty-second conference* (p. 484). Helsinki: ILA.
International Law Commission (ILC). (1994). Draft articles on non-navigational uses on international watercourses.In *Report of the international law commission on the work of its forty-sixth session*. UN Gaor 49th Sess., Suppl. No. 10, UN Doc. A/49/10.
International Law Commission (ILC). (2001). Draft articles on the prevention of transboundary harm from hazardous activities. In *Report of the fifty-third session, yearbook of the international law commission*, 2001, vol. II, Part Two.
Jennings, R. (1986). Equity and equidistance principles. *Annuaire Suisse de Droit International, 27*, 407.
Jimenez de Arechaga, E. (1960). International legal rules governing the use of waters of international watercourses. *2 Inter-American Law Review, 329*.
Kroes, M. (1997). The protection of international watercourses as sources of fresh water in the interest of future generations. In E. H. P. Brans, E. J. de Haan, J. Rinzema, & A. Nollkaemper (Eds.), *The scarcity of water: Emerging legal and policy responses*. The Hague: Kluwer Law International.
Lapidoth, R. (1987). Equity in international law. *22 Israel Law Review, 161*.
Lipper, J. (1967). Equitable utilization. In A. H. Garretson et al. (Eds.), *The law of international drainage basins* (pp. 16–26). New York: Dobbs Ferry/Oceana Publications.
Lowe, V. (1992). The role of equity in international law. *Australian Yearbook of International Law, 12*, 54–81.
Maljean-Dubois, S. (2011). The making of international law challenging environmental protection. In Y. Kerbrat & S. Maljean-Dubois (Eds.), *The transformation of international environmental law* (pp. 25–54). Paris: Pedone & Hart.
McCaffrey, S. C. (2001). *The law of international watercourses: Non-navigational uses*. Oxford: Oxford University Press.
McCrae, D. (1981). Proportionality and the Gulf of main boundary dispute. *Canadian Yearbook of International Law, 19*, 287–302.
McIntyre, O. (2004). The emergence of an "ecosystems approach" to the protection of international watercourses under international law. *Review of European Community and International Environmental Law, 13/1*, 1.
McIntyre, O. (2007). *Environmental protection of international watercourses under international law*. Farnham: Ashgate.
McIntyre, O. (2010). The proceduralization and growing maturity of international water law: *Case concerning pulp mills on the river Uruguay (Argentina v. Uruguay). Journal of Environmental Law, 22*(3), 475–497.
McIntyre, O. (2011). The world court's ongoing contribution to international water law: The *pulp mills* case between Argentina and Uruguay. *Water Alternatives, 4*(2), 124–144.
McIntyre, O. (2013a). Utilization of shared international freshwater resources – The meaning and role of "equity" in international water law. *Water International, 38*(2), 112–129.

McIntyre, O. (2013b). The contribution of procedural rules to the environmental protection of transboundary rivers. In L. Boisson de Chazournes, C. Leb, & M. Tignino (Eds.), *Freshwater and international law: The multiple challenges* (pp. 239–265). Cheltenham: Edward Elgar.
McIntyre, O. (2014). The protection of freshwater ecosystems revisited: Towards a common understanding of the "ecosystems approach" to the protection of transboundary water resources under international law. *Review of European, Comparative and International Law, 23*(1), 88–95.
McIntyre, O. (2015). The UNECE water convention and the human right to access to water: The protocol on water and health. In A. Tanzi, O. McIntyre, A. Kolliopoulos, A. Rieu-Clarke, & R. Kinna (Eds.), *The UNECE convention on the protection and use of transboundary watercourses and international lakes: Its contribution to international water cooperation* (pp. 345–366). Brill Nijhoff: Leiden.
Nelson, L. D. M. (1990). The role of equity in the delimitation of maritime boundaries. *American Journal of International Law, 84*(4), 837–858.
Rothpfeffer, T. (1972). Equity in the North Sea continental shelf cases. *Nordisk Tidsskrift for International Ret, 42*, 81–137.
Sands, P., & Peel, J. (2012). *Principles of international environmental law* (3rd ed.). Cambridge: Cambridge University Press.
Schmeier, S. (2013). *Governing international watercourses: River basin organizations and the sustainable governance of internationally shared rivers and lakes*. Abingdon: Routledge.
Tanzi, A., & Arcari, M. (2001). *The United Nations convention on the law of international watercourses*. The Hague: Kluwer Law International.
Tully, S. R. (2003). The contribution of human rights to freshwater resource management. *Yearbook of International Environmental Law, 14*, 101.
UNFCCC. (1992). *United Nations framework convention on climate change*. 31 ILM 849.
UN-INWEH. (2013). *Water security and the global water Agenda: A UN-water analytical brief.* Hamilton: United Nations University.
United Nations Conference on Environment and Development (UNCED). (1992). *Rio declaration on environment and development*. UN Doc. A/CONF.151/26 (vol. I)/31 ILM 874 (1992).
United Nations Economic Commission for Europe (UNECE). (1992). Convention on the protection of transboundary watercourses and international lakes. *International Legal Materials, 312*, 1312.
United Nations Economic Commission for Europe (UNECE). (2004). *UNECE water convention MoP Decision III/I, 'amendment to articles 25 and 26 of the convention', UN Doc. ECE/MP/WAT/14 (12 January 2004)*. Adopted 28 November 2003, entered into force 6 February 2013.
United Nations Environment Programme (UNEP). (1978, May 19). *Draft principles on conduct in the field of the environment for guidance of States in the conservation and harmonious utilization of natural resources shared by two or more States*. UNEP Governing Council Decision 6/14.
United Nations, Report of the United Nations Conference on the Human Environment (UNCHE). (1972). Stockholm 5-16 June 1972 (UN Publication Sales No. E.73.II.A.14).
UN-Water. (2015). *Methodological note: Proposed indicator framework for monitoring SDG targets on drinking water, sanitation, hygiene and wastewater*. Geneva: WHO/UNICEF.
UNWC. (1997) *United Nations convention on the law of non-navigational uses of international watercourses*. 36 ILM 700.
Wouters, P. K., & Rieu-Clarke, A. (2001). The role of international water law in promoting sustainable development. *Water Law, 12*, 281.
Yihdego, Z., & Rieu-Clarke, A. (2016). An exploration of fairness in international law through the Blue Nile and GERD. *Water International, 41*(4), 528–549.

Chapter 4
Water as a Human Right in the Global South: Ethical, Legal and Sociopolitical Dimensions

Patricia Avila-García

Abstract This chapter seeks to elucidate the scope and limitations of the international recognition of the human right to water, and the course that States must follow in order to fulfill the commitments they have assumed. It analyzes the links among the human right to water, water security, and environmental justice, given that the existence of a water crisis that requires a human-oriented approach and solution has been recognized. The chapter is divided into three parts: the first discusses the water crisis and its relation to water security and the human right to water, while the second focuses on the ethical and legal dimensions of the human right to water and postures that favor, or oppose, its recognition. Finally, the third part focuses on the sociopolitical dimension of the human right to water; that is, the role of the State in complying with and/or violating this right, and the defensive actions taken by civil society in Latin American countries (*i.e.*, those in the Global South).

Keywords Human right to water • Water security • Environmental justice • State • Latin America

4.1 Introduction

Important advances in the recognition of human rights have been made internationally from the mid-twentieth century to the present, in particular regarding rights considered second- and third-generation, which include the conception of water as essential for life, human wellbeing, food production and development. However, it was only with the political recognition of the social and environmental impacts of the dominant economic model –*i.e.*, the water crisis– that advances could be made in conceiving the right to water as expressed in General Comment 15 of the United Nations' International Covenant on Economic, Social and Cultural Rights, approved

P. Avila-García (✉)
Research Institute in Ecosystems and Sustainability (UNAM-IIES), National Autonomous University of Mexico, Mexico City, Mexico
e-mail: pavila@cieco.unam.mx

D. Devlaeminck et al. (eds.), *The Human Face of Water Security*, Water Security in a New World, DOI 10.1007/978-3-319-50161-1_4

in 2002. That recognition derived from the idea of development, which implies a more comprehensive perspective of the right to water, one that protects individuals, communities and social groups. This notion is not restricted to supplies of drinking water and sanitation issues, but also recognizes the importance of water for food production, the functioning of ecosystems and peoples' development.

In addition, it is important to highlight the role of social movements and civil organizations in the defense of water from the international pressures of transnational organisms like the World Bank, which conceive water as an economic asset valued as a commodity, and so seek to advance towards its privatization and the assignment of property rights to end-users (eg. households, services, agriculture, industry). Particularly noteworthy in the movement to have water recognized as a human right was the leadership of the Bolivian government (in the aftermath of the social struggles that opposed the privatization of water in Cochabamba and Altos-La Paz) at the Fourth World Water Forum held in Mexico in 2006, and later at the United Nations (UN). The resolution was finally approved by the UN's General Assembly in July 2010; thus recognizing the rights to drinking water and sanitation as essential for the full enjoyment of life and the exercise of all other human rights. Nations and international organizations were summoned to provide financial resources and propitiate increased technological capacity and transference in and to developing countries to provide populations with affordable access to drinking water and sanitation.

At the international level, this constitutes a significant improvement because signing countries were formally committed to adopting concrete, gradual measures to make the human right to water a reality. To achieve this goal, nations must design strategies that include: the necessary legal reforms of their constitutions and corresponding water legislation to recognize this right; the application of public policies and programs to provide drinking water and sanitation; and the assignment of technological and financial resources for the projects and infrastructure required. Once ratified, compliance is compulsory and no backward steps are accepted except in extraordinary situations, such as economic crises or political instability, when States that require it may receive international aid. Moreover, violations of water rights through omission or commission must be processed judicially to impose admonishments or sanctions for non-compliance.

In light of the foregoing, this chapter seeks to elucidate the scope and limitations of the international recognition of the human right to water, and the course that States must follow in order to fulfill the commitments they have assumed. It analyzes the links among the human right to water, water security, and environmental justice, given that the existence of a water crisis that requires a human-oriented approach and solution has been recognized. The chapter is divided into three parts: the first discusses the water crisis and its relation to water security and the human right to water, while the second focuses on the ethical and legal dimensions of the human right to water and postures that favor, or oppose, its recognition. Finally, the third part focuses on the sociopolitical dimension of the human right to water; that is, the role of the State in complying with and/or violating this right, and the defensive actions taken by civil society in Latin American countries (*i.e.*, those in the Global South).[1]

4.2 The Water Crisis, Water Security and Human Rights

The water crisis is associated with processes of urbanization and demographic dynamism, as well as economic growth and industrialization, all of which generate greater demands on water and have profound repercussions on the availability and quality of both surface water and groundwater. It is also associated with social inequality in rural and urban areas that leads to limited access to drinking water and sanitation, especially for the poorest and most marginalized sectors of the population.

But this crisis also involves more intense competition for the use of water, which generates conflicts over control among social, economic and political stakeholders, all of whom have different valuations of this resource. Some sectors see water as a shared asset of high social, cultural and ecological value; others perceive it as an economic good (commodity) that is subject to private ownership and market laws, while still others see it as a political resource, a source of power and control. In many cases, such conflicts cross national borders to generate diplomatic and political tensions at the international level.

Complicating this scenario is the reality of environmental degradation (deforestation, water pollution), the loss of ecosystem services related to water (recharge of aquifers, flood control), alterations in patterns of precipitation, and the occurrence of extreme events (droughts, flooding) due to global climate change. Therefore, increasing vulnerability of both populations and ecosystems can be foreseen as a consequence of climate change, since there will be greater uncertainty in water availability, alterations of agricultural cycles, and irreversible ecological degradation.[2]

But this is also associated with deficient governance, revealed in the incapacity of States to resolve the demands and conflicts over water that involve populations and economic activities, from the local to the transnational level. This can be explained by the weakness of existing juridical and institutional frameworks, and by financial and technological restrictions that impede the adequate management of this resource.

In response to the water crisis associated with such global processes as climate change, economic globalization, urbanization and governance, several visions have

[1] The Global South largely corresponds to the Third World: defined as the poorer, less-developed region of the world.

[2] According to UN Water (2012: 7): "as water demand and availability become more uncertain, all societies become more vulnerable to a wide range of risks associated with inadequate water supply, including hunger and thirst, high rates of disease and death, lost productivity and economic crises, and degraded ecosystems. These impacts elevate water to a crisis of global concern."

emerged that attempt to approach this as an issue of national and international security, since human populations, ecosystems and life itself are at risk. In this regard, the Ministerial Declaration of the Second World Water Forum (2000) declared the importance of achieving water security in the face of the looming water crisis through advances designed to overcome the following challenges:

- *"Meeting basic needs: to recognize that access to safe and sufficient water and sanitation are basic human needs and are essential to health and well-being, and to empower people, especially women, through a participatory process of water management.*
- *Securing the food supply: to enhance food security, particularly of the poor and vulnerable, through the more efficient mobilization and use, and the more equitable allocation of water for food production.*
- *Protecting ecosystems: to ensure the integrity of ecosystems through sustainable water resources management.*
- *Sharing water resources: to promote peaceful co-operation and develop synergies between different uses of water at all levels, whenever possible, within and, in the case of boundary and trans-boundary water resources, between states concerned, through sustainable river basin management or other appropriate approaches.*
- *Managing risks: to provide security from floods, droughts, pollution and other water-related hazards.*
- *Valuing water: to manage water in a way that reflects its economic, social, environmental and cultural values for all its uses, and to move towards pricing water services to reflect the cost of their provision. This approach should take account of the need for equity and the basic needs of the poor and the vulnerable.*
- *Governing water wisely: to ensure good governance, so that the involvement of the public and the interests of all stakeholders are included in the management of water resources" (Second World Water Forum* 2000*).*

In recent years, the notion of water security has been expanded. Here, the concept developed by United Nations Water (UN Water) is particularly relevant in terms of achieving sustainable management of the water cycle to benefit both human wellbeing and ecosystem preservation, because it offers a comprehensive, holistic formulation that links two key dimensions: one human (sustenance, wellbeing, socioeconomic development), the other biophysical (water quality, extreme climatic events, ecosystems). In other words, it entails satisfying basic needs related to water supplies, protecting water quality from pollution, and managing conflicts from the individual to group levels, and from the local to international scales. UN Water (2013) defines water security as:

> …the capacity of a population to safeguard sustainable access to adequate quantities of acceptable quality water for sustaining livelihoods, human well-being, and socio-economic development, for ensuring protection against water-borne pollution and water-related disasters, and for preserving ecosystems in a climate of peace and political stability. This definition implies that water is managed sustainably throughout the water cycle and is done so through an inter-disciplinary focus, so that it contributes to socio-economic development

> and reinforces societal resilience to environmental impacts and water-borne diseases without compromising the present and future health of populations and ecosystems. Achieving water security requires allocation among users to be fair, efficient and transparent; that water to satisfy basic human needs is accessible to all at an affordable cost to the user; that water throughout the water cycle is collected and treated to prevent pollution and disease; and that fair, accessible and effective mechanisms exist to manage or address disputes or conflicts that may arise. The concept operates at all levels, from individual, household and community, to local, sub-national, national, regional and international settings, and takes into account the variability of water availability over time. (UN Water 2013: 1).

This requires recognizing the fact that water, ecosystems and society are closely-linked, and the importance of protecting the biophysical environment and populations to ensure sustainable development, such that the rights of both humans and nature herself must be protected for the sake of present and future generations.

From an ethical standpoint, water security is linked to the human right to water when emphasis is placed on protecting individuals and groups (and, hence, also human dignity), satisfying essential needs, and guaranteeing water supplies through the sustainable management of ecosystems. In this regard, water security becomes feasible to the extent that countries protect both nature's and society's rights to water, harmonize economic, social and environmental objectives, reorient production and consumption patterns to adhere to a rationality of sustainability, and foster solidarity through international cooperation and compliance based on agreements of global importance (eg. regarding climate change).

It is important to mention that recognizing the human right to water does not guarantee, in and of itself, a solution to the water crisis, nor greater water security in individual nations (UN Water 2013), because the problems associated with water are complex and by no means restricted to a purely legal dimension (eg. social inequalities, governance and globalization).

> The formal recognition of a human right to water and sanitation will not in itself alter the realities on the ground, such as water scarcity, polluted wells and rivers, poor governance, a lack of investments in infrastructure, or the prevalence of inequalities. Nevertheless, it has already generated political will, providing a framework for development, conflict resolution, and accountability in the water services sector. (UN Water 2013: 12)

In relation to this we know of several contrasting cases: local initiatives that contribute to achieving water security, though with no explicit recognition of the human right to water; and citizen-based experiences of applying this right that ensure water security by protecting the requirements of the population (Argentina and South Africa, among others). But there are also examples of ecosystem deterioration when priority is given to the demand for water by urban populations and the development of economic activities (eg. irrigation, mining, industry) that affect water security in transboundary regions and watersheds (eg. construction of hydroelectric dams and water transfers between basins).

For purposes of this section, we exemplify only the first case by discussing how one indigenous region in Mexico characterized by high levels of poverty and a severe scarcity of water, achieved greater water security through a process of social participation and organization by the population with the support of NGOs.

The "Water Forever Program" (*Programa Agua para Siempre*) introduced appropriate technologies into a sociocultural and ecological context in the indigenous Mixtec region that led to the restoration of severely degraded watersheds and the generation and harvesting of water to improve supplies for the population and food production for household consumption. This was achieved during a period in which the Mexican State had not yet included the human right to water in its Constitution. The actions performed by this Program contributed to greater water security in the region by satisfying social needs while also respecting the human dignity of the indigenous population.

4.3 The Human Right to Water: Ethical and Legal Dimensions

4.3.1 Ethical Dimension

According to Chociej and Adeel (2012: 123), human rights belong to a class of moral rights for they represent the minimum standards of acceptable treatment to which individuals are naturally entitled: "These rights can be taken up within legal frameworks, as legal rights. When this happens, issues of fundamental moral concern become translated into a paradigm where citizens become rights-holders and governments become duty-bearers."

It is in this sense that the ethical dimension becomes central when referring to human rights, because it implies recognition of all aspects of human dignity; *i.e.*, the rights to life, wellbeing, freedom, self-determination, and development. This includes both individuals and groups since the fundamental human rights are moral rights that take precedence over legal ones (Murillo 2014).

These fundamental rights are born of social demands and struggles for their effective vindication as guarantees that have been sustained throughout history by ethical values (eg. struggles against slavery and colonialism that seek the freedom and self-determination of peoples). Also, the great tragedies that humanity suffered during the twentieth century (world wars, genocides, famines) propitiated transit from a moral to a legal recognition of human rights through binding international agreements that oblige States to incorporate them into their legal frameworks, policies, and national programs. The Universal Declaration of Human Rights proclaimed by the United Nations in 1948 followed these guidelines.

In the case of water, the ethical dimension is associated with human dignity and the right to life, given that the availability of water is a basic need of all individuals and groups. Access to water of adequate quality and in sufficient amounts will make it possible to guarantee human health and wellbeing. In this regard, Gleick (1999) states that the correct approach to the human right to water is to satisfy the basic needs for domestic uses (consumption, personal hygiene, food preparation). Therefore, to lead a dignified life, every person must be included in this basic condition of a guaranteed minimal availability of water, which means that excluding individuals and groups from this essential right is morally unacceptable, whether it be due to economic, political or cultural factors. Chociej and Adeel (2012), meanwhile, posit that the right to water is, by definition, a welfare right, while many other authors have argued that access to water is a pre-condition for human dignity (eg. Resolutions of the UN General Assembly and the UN Human Rights Council state this premise explicitly).

Furthermore, Jennings et al. (2009) propose seven ethical principles that must be included in water management: equal respect for human dignity; equity; proportionality; solidarity; the common good; a right relationship or responsible stewardship; and inclusive, deliberative participation. These authors state that achieving human dignity must focus on satisfying the basic needs of promoting human health and wellbeing, a concept closely-linked to human rights.

Similarly, the ethical notion of water leads us to consider social equity and justice in the sense that the human right to water must be ensured for all individuals, especially the poorest and most marginalized sectors in economic, cultural and political terms. This entails international recognition of the fact that millions of people around the world lack the essential requirements of water to satisfy their needs. In addition, this notion justifies the international aid and cooperation that are required, while exhibiting the responsibility of States to resolve a problem that affects human dignity in their respective countries.

It is important to note that the ethical notion of satisfying essential needs must take into account not only present but also future generations, since water is a finite resource that demands adequate conservation and management. Added to this is the high vulnerability of water to global climate change. For these reasons, recognizing the environmental dimension of water and nature is essential for human survival. In the absence of adequate water quality and quantity it is impossible to satisfy the population's requirements for wellbeing and a dignified life. Thus, conservation of ecosystems and ecosystem services related to water are of crucial importance for humanity.

In this regard, there is a case of environmental ethics that is especially interesting as it goes beyond a purely anthropocentric position: the Law of the Rights of Mother Earth in Bolivia. Legislation that defines Nature as a collective subject of public interest, and declares both Her and life-systems (which combine human communities and ecosystems) as titleholders of the inherent rights stipulated in law. Specifically, it establishes the right to preserve the quality and composition of water to sustain life systems; to protect it from contamination; and to renew the life of Mother Earth and all its components (Plurinational State of Bolivia 2010).

Also on the side of environmental justice are the so-called ethical tribunals (eg. the International War Crimes Tribunal or Russell-Sartre Tribunal and the People's Permanent Tribunal). These are spaces created by society where States prove incapable –through omission or commission– of safeguarding the common good and the interests of individuals and communities. Particularly important is the case of The Latin American Water Tribunal, created in 2000 as an autonomous international body that promotes environmental justice by helping to resolve conflicts involving water in the region. It is based on the principles of co-existence, respect for human dignity, solidarity among peoples, environmental responsibility, and the sacredness of living forms. Its role is essentially didactic as it seeks to raise consciousness and strives to achieve a political and social consensus to transform ethical-environmental values and effectuate change in the dominant paradigms (Avila 2007; Bogantes 2007).

Tribunals of consciousness depend on the strength of moral condemnation and civil demonstrations that defend the fundamental right of Latin Americans to water of suitable quantity and quality. People have a right, above those of corporate and State powers, to the use and protection of their water resources for generations both present and future. Society must make a conscious decision concerning the importance of managing water to ensure social and environmental sustainability, while remaining alert to public or private projects that currently or potentially affect water systems in Latin America (Bogantes 2007).

4.3.2 The Legal Dimension

Human rights are essential for the development of individuals and constitute an attempt to respond to failures in the economic and political systems responsible for causing unacceptable human misery. In legal terms, these rights are universal, irrevocable, inalienable, indefeasible, undividable, and interdependent, because they share a common status and hierarchy and are mutually complementary.

The terms used to refer to human rights have varied: individual guarantees; natural rights; fundamental rights; economic, social and cultural rights; and political rights, but a distinction is often made between the concept of human rights and individual guarantees in which the former are general, abstract notions, and the latter individualized, concrete ideas that seek to support and protect people.

The Universal Declaration of Human Rights, a non-binding legal document adopted in 1948, was made legally-binding through two separates treaties in 1966: the International Covenant on Economic, Social and Cultural Rights (ICESCR), and the International Covenant on Civil and Political Rights (ICCPR). These international instruments not only define economic, social, cultural, civil and political rights, but also establish general principles, criteria and standards.

Human rights are classified in three generations based on the historical context in which they emerged and their social and formal recognition (Sandoval 2001):

(a) The first generation includes civil and political rights recognized in the second half of the eighteenth century that incorporated the ideals of the French Revolution and U.S. independence, including equality before the law and freedom of thought, conscience and religion.
(b) The second generation embraces economic, social and cultural rights conquered through the labor movement, nineteenth-century socialist ideas, and the social struggles of the first half of the twentieth century (eg. the 1910 Mexican

Revolution and the 1917 Russian Revolution). They specify the rights to education, food, housing and employment.

(c) The third generation corresponds to rights that came to be recognized during crucial events that marked the second half of the twentieth century: the internationalization of conflicts, national liberation movements, the non-aligned countries movement during the Cold War, and environmental degradation. They include the right to peace, self-determination, development, and a healthy environment.

Significantly, the right to water is included in all three generations, since in conditions of inequality it is impossible to guarantee a population's access to water, just as it is impossible to produce sufficient food or supply dignified housing. The same occurs when ecosystems are degraded, as it becomes impossible to enjoy the services they provide in relation to water.

As a result, the notion of the right to water is implied in international legal instruments that are binding on States, including the Geneva Convention for Humanitarian Treatment of Civilians during Wars (1949), the International Covenant on Economic, Social and Cultural Rights (1966), the Committee on the Elimination of Discrimination against Women (1979), and the Convention on the Rights of the Child (1989). Particularly noteworthy is General Comment 15 of the ICESCR (2002) which attempts to specify the human right to water (UN Economic and Social Council 2002; Scanlon et al. 2004).

In addition, non-binding international declarations and programs related to water have been issued since the 1970s: the Mar del Plata Action Plan on Water Development and Management that was adopted at the UN Water Conference in March 1977; Agenda 21 and the Rio Declaration on the Environment and Development, adopted at the UN Conference on the Environment and Development in June 1992; and the Habitat Agenda, adopted at the second UN Conference on Human Settlements in June 1996.

It is important to note that by signing and ratifying such International Resolutions, States are compelled to gradually introduce the legal and institutional changes necessary to design and apply public policies that propitiate compliance with this right in their countries. In this regard, the failure to implement change is not treated as a violation of human rights but, rather, as a sign that the State needs to be encouraged and exhorted to take actions to ensure compliance. At the same time, once these resolutions and their associated commentaries have been signed and ratified, any regression in respect for the rights stipulated therein is considered unacceptable (Gutierrez 2007).

4.3.2.1 General Comment 15 of the International Covenant on Economic, Social and Cultural Rights Regarding Water

The right to water was recognized in Articles 11 and 12 of the ICESCR, a multilateral treaty that took effect in 1976 and was ratified by 160 countries. To advance this project and provide support to the ICESCR (through comments), the UN appointed a Committee on Economic, Social and Cultural Rights (CESCR) made up of members of State parties. The Committee's comments, however, only constitute a non-legal, non-binding regulatory framework that –it is hoped– will be applied progressively in each country (Langford and Khalfan 2007;UN Development Program 2006).

In 2002, the CESCR issued General Comment 15, which confirmed the right to water in the following terms: "Water is a limited natural resource and a public good fundamental for life and health. The human right to water is indispensable for leading a life in human dignity. It is a prerequisite for the realization of other human rights" (UN Economic and Social Council 2002).

Despite the reluctance of countries such as the United States, Australia and Saudi Arabia to raise water to the category of a human right, General Comment 15 was accepted by nearly 70 countries in 2003. The European Parliament declared water as a human right and several nations, including the Netherlands and Great Britain, reformed their legislation. In the case of Latin America, Uruguay was the first country to approve a referendum adopting water as a human right and to reform its constitution accordingly (Langford and Khalfan 2007).

General Comment 15 States that water must be treated as a social and cultural good rather than as an economic commodity, and that the exercise of the right to water must be sustainable so that future generations may also enjoy it. However, the notion of the human right to water was simplified to make it more acceptable to most States: "The human right to water entitles everyone to sufficient, safe, acceptable, physically accessible and affordable water for personal and domestic uses. An adequate amount of safe water is necessary to prevent death from dehydration, to reduce the risk of water-related disease and to provide for consumption, cooking, personal and domestic hygienic requirements" (UN Economic and Social Council 2002).

In addition, General Comment 15 establishes a series of normative aspects that must be complied with to guarantee the human right to water (see Box 4.1).

4.3.2.2 Resolutions of the United Nations General Assembly and the Council on Human Rights Recognizing the Human Right to Drinking Water and Sanitation

According to Langford and Khalfan (2007), the approach to water as a human right is based on universally-valid legal and normative principles that should lead to: (a) governments giving priority to access to water, especially for the poorest and most vulnerable sectors of the population; (b) assuming water supplies as a right rather than as an act of charity or a commodity; (c) assuring that water supplies do not generate discrimination due to socioeconomic status, culture, race, gender, religious belief, political affiliation, or ideology; (d) consultations with, and participation by, society in decision-making processes, especially regarding access to water, supply systems and conservation; and, (e) greater responsibility by national governments, the international community and the private sector to guarantee access to water.

Box 4.1 General Comment No. 15. Normative Content Right to Water
General Comment No. 15 establishes a series of normative aspects of the human right to water, including:

- "the right to maintain access to existing water supplies necessary to the right to water,"
- "the right to be free from interference," including "arbitrary disconnections or contamination,"
- The entitlement to a "system of water supply and management which provides equality of opportunity" for the enjoyment of the right to water,
- Quality and quantity should be determined by "volumetric quantities and technologies", but "treated as a social and culture good, and not primarily as an economic good,"
- The attainment of the right to water must be conducted in a sustainable way, "ensuring the right can be realized for present and future generations."

Furthermore, General Comment No. 15 states that, while "the adequacy of water required for the right to water may vary according to different conditions, the following factors apply in all circumstances," including:

- *Availability*: Water supply for individuals must be "sufficient and continuous for personal and domestic uses", including drinking, personal sanitation, washing clothes, food preparation, as well as personal and household hygiene. The quantity should meet WHO guidelines and certain groups or individuals may need more water in consideration of work, climate and health conditions.
- *Quality*: Water must be safe, defining "safe" as "free from micro-organisms, chemical substances and radiological hazards that constitute a threat to a person's health." It must also be of an "acceptable colour, odour and taste."
- *Accessibility*: "Water and water facilities and services have to be accessible to everyone without discrimination," where accessibility is based on four dimensions, including: physical accessibility, economic accessibility, non-discrimination and information accessibility.

Source: UN Economic and Social Council (2002).

From an integrated perspective, the human right to water is considered a necessary condition for achieving adequate standards of living, since it is essential for survival. However, in order to achieve total enforcement and respect for this right, other human rights must be equally guaranteed; such as the right to health, housing and nourishment; to life and dignity; to freedom from discrimination; to participation; to personal and community integrity; and to development

(UN Development Program 2006; UN Economic and Social Council 2002). In other words:

(a) Without adequate amounts and quality of water no population can be healthy nor can human life and dignity be respected.
(b) Without water supplies in localities and houses inhabited by deprived people no guarantee exists of the volume required for survival and discrimination, and social exclusion will be fomented.
(c) Without water available for food production the elementary dietary requirements of the population cannot be satisfied and self-sufficiency in food becomes vulnerable.
(d) Without social participation in decision-making regarding access, use and management of water, and without respect for personal and communitarian integrities, the emergence of conflicts will be exacerbated, threatening both human life itself and harmony among peoples.
(e) Without respect for the environment and development it will be impossible to ensure water availability to future generations.

In order to advance towards the recognition of the human right to water as a binding obligation for States, in 2006 the UN Sub-Commission on the Promotion and Protection of Human Rights approved guidelines for the fulfillment of the human right to drinking water and sanitation. Also, the UN Development Program acknowledged the importance of recognizing the right to water as a fundamental human right. In 2008, the Council on Human Rights created the mandate of an "Independent expert on the issue of the obligation of human rights related to access to safe drinking water and sanitation" to clarify the scope and content of such obligations (UN World Heath Organization 2010).

Later, reports were made to the High Commissioner for Human Rights on issues of human rights related to drinking water and sanitation, as well as by the Independent expert on the question of the obligatory nature of human rights related to access to drinking water and sanitation. Such actions derived in the explicit recognition of the human right to drinking water and sanitation through the binding resolution adopted by the UN Assembly in July 2010 that is obligatory for all signing States (UN General Assembly 2010).

Here, we must mention the role played by certain international NGOs and grassroots organizations that pressured for the explicit recognition of the human right to water (eg. Declarations at the Alternative World Water Forums in Mexico, 2006 and Istanbul, 2009); as well as the political lobbying by the Plurinational State of Bolivia, which proposed the wording for the international resolution with the backing of 33 other UN members. That measure received 122 votes in favor, and zero votes against, though 41 countries abstained (including the U.S., Canada and the UK).

In Pardy's words (2011: 907): "…delegates from abstaining countries said that consensus was lacking, that the declaration was premature and in the wrong forum, and that the meaning of such a right in international law was uncertain.." Also Murthy (2013: 103) states: "The abstaining states, such as the United States and Canada, may have been concerned that the right to water and sanitation was not explicitly tied to rights recognized in the ICESCR. As a result, the General Assembly resolution could be interpreted as creating new rights. Moreover, the General Assembly resolution was silent on the role of non-state actors and privatization.."

This resolution is based on ethical principles that include: recognizing that millions of people around the world do not have access to drinking water and basic sanitation; accepting that these two requirements must be available equitably to achieve overall human rights; the responsibility of States to promote and protect all human rights and treat them in a comprehensive, fair and equitable manner; and the commitment of countries to achieve the Millennium Development Goals regarding the application of measures that reduce the deficits related to drinking water and sanitation. In summary, the resolution has three main points (UN General Assembly 2010: 2–3):

- *"Recognizes the right to safe and clean drinking water and sanitation as a human right that is essential for the full enjoyment of life and all human rights;*
- *Calls upon States and international organizations to provide financial resources, capacity-building and technology transfer, through international assistance and cooperation, in particular to developing countries, in order to scale up efforts to provide safe, clean, accessible and affordable drinking water and sanitation for all;*
- *Welcomes the decision by the Human Rights Council to request that the independent expert on human rights obligations related to access to safe drinking water and sanitation submit an annual report to the General Assembly, and encourages her to continue working on all aspects of her mandate and, in consultation with all relevant United Nations agencies, funds and programmes, to include in her report to the Assembly, at its sixty-sixth session, the principal challenges related to the realization of the human right to safe and clean drinking water and sanitation and their impact on the achievement of the Millennium Development Goals."*

This resolution was expanded by the Council on Human Rights (UN General Assembly 2011) to recognize that the human right to drinking water and sanitation is derived from the right to adequate living standards, and is closely-related to achieving physical and mental health, and to the rights to life and human dignity. In short, it ratifies essential ethical principles concerning the human dimension of the right to water, while emphasizing the need to focus on local and national perspectives in considering this issue, setting aside questions of international watercourse law and all trans-boundary water issues. It also posits a gradual realization of this right: "States have the primary responsibility to ensure the full realization of all human rights, and must take steps, nationally and through international assistance and cooperation, especially economic and technical, to the maximum of its available resources, to achieve progressively the full realization of the right to safe drinking water and sanitation by all appropriate means, including particularly the adoption of legislative measures in the implementation of their human rights obligations" (UN General Assembly 2011: 3).

Since then, the UN and other international forums have issued additional resolutions designed to strengthen the concept and content of the human right to water in international jurisprudence. Of particular note are the declarations emitted by the UN Human Rights Council in 2012, 2013 and 2014, and by the General Assembly in 2013. Meanwhile the ministerial declaration of the Sixth World Water Forum, held in Marseilles in 2012, finally recognized the human right to water after attempts at two earlier forums had failed (Mora and Dubois 2015).

For example, the UN Human Rights Council's Resolution dated September 2012 reaffirmed the primordial responsibility of States to guarantee the full effectiveness of all human rights and to implement the economic and technical measures at the national level and through international cooperation required to progressively transform the established right to drinking water and sanitation into a concrete reality. This will demand taking legislative action to fulfill all obligations related to questions of human rights (Mora and Dubois 2015).

In turn, the Resolution of December 2013 of the UN General Assembly stressed the importance of guaranteeing the gradual realization of the human right to drinking water and sanitation in a non-discriminatory manner by progressively eliminating inequalities of access that affect groups which are vulnerable and marginalized for reasons of race, sex, age, incapacity, ethnic origin, culture, religion or national or social origin, or for any other motive (Mora and Dubois 2015).

In legal terms, recognition of the human right to water and sanitation has profound implications for international law due to the many juridical resolutions and instruments that have been emitted by the UN's multilateral organisms and international conferences and summit meetings. As a result of these developments, the human right to water and sanitation has been incorporated into the system of international law as an enforceable right with a binding character; that is to say that countries are obliged to respect it and to generate the legal and political frameworks necessary for its full realization. Moreover, since they form part of international law, individual States have the obligation to respect, protect and apply all human rights within their territories. At present, the tendency is to work towards the universal recognition of this right through gradual changes in each country's legislation, investments in infrastructure for water supplies and sanitation systems, improvements in jurisprudence to better attend to cases of non-compliance or violations, and greater international cooperation between countries in the Global North and South.

Some authors, such as Pardy (2011, 2012) and Leb (2012), have pointed out that this recognition does not resolve, in the strict sense of this term, problems related to water but, rather, exacerbates others and generates new ones. For example, at the international level countries that suffer water stress could demand support from those that have an excess of water; nations in the Global South could demand greater financial and technological cooperation from countries in the Global North; and in countries that share transboundary basins in which water flows from higher elevations to lower ones, the latter could be affected by measures taken by the former as it seeks to establish this right from a purely local focus. But difficulties could also emerge on the national plane if priority is given to supplying human populations with water by constructing systems that damage ecosystems; if subsidies and other forms of support related to water are given to population sectors that are not necessarily the poorest or most marginalized; or if States receive demands for water that they cannot satisfy due, for example, to drought or low availability. In this regard, Leb (2012) proposes the need for significant advances in international jurisprudence, including treaties and agreements that clearly define such issues in order to improve water management within and between countries.

Bakker (2007) and Murthy (2013) observe that opposition to the privatization and commodification of water have dominated the discourse of the NGO's and grassroots organizations that promote recognition of water as a basic human right. However, they suggest that this approach to the problem is poorly-framed, because the form of water management (*i.e.*, public *vs.* private) does not determine access to water. There are cases in which private water management –well-regulated by the State– guarantees an adequate service; and others where public management fails to comply with the minimum standards of water quality and quantity for the population. In this regard, Pardy (2011: 909) sustains that:

> The dark irony of international water rights is that they could frustrate the very objectives they are intended to achieve. The ideology underlying the campaign for water rights contains two conflicting premises. The first is that governments cannot be trusted to make clean water available. Therefore, norms of international law must be brought to bear upon them. An international right is the means whereby national governments can be held accountable. The second premise is that only governments can be trusted to deal with water, and certainly the private sector cannot. The nature of the proposed rights implies that only governments may provide water, and therefore must do so in the form of water monopolies.

From a positive perspective, recognition of water as a human right constitutes a significant advance to the extent that individual States are capable of generating mechanisms to resolve conflicts, reorient priorities and invest to satisfy the basic needs of populations, and to promote international collaboration and the participation of diverse actors in solving problems related to water. In addition, it is a measure that can constrain the voracity of some actors who seek to commercialize this resource by selling bottled water and privatizing water-supply services, actions that could restrict access by the poorest sectors of the population because of high costs, as occurred in the extreme case of the so-called "water wars" in Bolivia.

Also, recognition enriches jurisprudence by providing formal legal stipulations that protect individuals and groups in cases where their essential human rights are affected, including the right to water (UN World Health Organization 2003). It represents a first step in legal terms, given that States are made responsible for applying and monitoring this right, so that human rights are effectively respected by States and other stakeholders, including transnational corporations. Civil society must also assure fulfillment of the right to water by participating in the consolidation of national legal frameworks, and in the decision-making processes involved in the design and application of public policies and programs regarding water.

4.4 The Human Right to Water: Sociopolitical Dimensions

The system of the United Nations have emitted recommendations that expand the content of the notion of the human right to water and specify the role that States must assume to gradually guarantee compliance at the national level. In this sense, some countries have effectuated changes in their laws and public policies to enforce this right. But there are other nations where violations of this human right by States and private actors have multiplied. This has propitiated the emergence of social movements that seek environmental justice, especially in the Global South like Latin America, where groups have appeared before national and international instances to demand fulfillment of this human right.

4.4.1 States and the Human Right to Drinking Water and Sanitation: A Proposal from the United Nations

As stated by the UN Development Program (UN Development Program 2006) in its Human Development Report, the main challenges of public policy are to provide substance to the notion of water as a human right, and to assume the principles of equity, universality and non-discrimination. Hence, exclusion from water supplies and sanitation based on socioeconomic level, group affiliation, or place of residence would all constitute violations of the human right to water.

According to the United Nations' World Health Organization (2010), due to discrimination, stigmatization, or both, some groups find it especially difficult to exercise their right to water. In order to efficiently protect this right, it is necessary to understand the concrete situation of individuals and groups, especially the most vulnerable ones. States must adopt the necessary positive measures to ensure that certain individuals or groups are not discriminated against through purpose or action. The most vulnerable groups include: the urban and rural poor, women, children, persons with disabilities, refugees and internally displaced persons, and indigenous peoples.

The UN World Health Organization (2010) reiterates the States' primary obligation to protect and promote human rights. These obligations are broadly defined and guaranteed by international human rights treaties that create binding obligations for the States that ratify them. The obligations of States fall into three categories*:*

(a) *"The obligation to respect requires States to refrain from interfering directly or indirectly with the enjoyment of the right to water.*
(b) *The obligation to protect requires States to prevent third parties from interfering with the right to water. States should adopt legislation or other measures to ensure that private actors comply with human rights standards related to the right to water.*
(c) *The obligation to fulfill requires States to adopt appropriate legislative, administrative, budgetary, judicial, promotional and other measures to fully realize the right to water. States must, among other things, adopt a national policy on water that: gives priority in water management to essential personal and domestic uses."*

The obligation of States to protect human rights includes ensuring that non-state actors do not infringe upon the right to water. Thus, other actors in society (individuals, intergovernmental and non-governmental organizations and businesses) all have responsibilities in promoting and protecting human rights (UN World Health Organization 2010).

In order to provide a context for the notion of the human right to water and to evaluate the advances made by each country, General Comment 15 of the ICESCR (UN Economic and Social Council 2002) proposed the criteria of availability, quality and accessibility, the latter sub-divided into several indicators (physical, economic, accessibility of information, and no discrimination).[3] These criteria are useful for defining the basic obligations of States regarding the human right to water, which include: guaranteed access to a quantity and quality of water that is sufficient and appropriate for personal and domestic uses and for preventing diseases; and ensuring the right of access to water and to water infrastructure and services without discrimination, above all, for vulnerable or marginal groups.

In cases of violation of the aforementioned criteria, or breaches of basic, specific obligations (as defined by each country), this Comment states that a distinction must be made between a State's inability to fulfill its obligations regarding the right to water and its reluctance to do so. In other words, violations may be due to acts of commission –*i.e.*, direct actions by the State or other entities in the absence of adequate regulation– or to acts of omission, which refer to the failure to adopt appropriate measures to guarantee full enjoyment of the universal right to water, to the absence of a national water policy, or to lapses in enforcing the applicable laws.

In consequence, the evaluation of compliance with international agreements regarding the design and application of public policies and the assignment of social spending to guarantee the human right to water for the poorest and most vulnerable population sectors (indigenous peoples, subsistence farmers, and residents of marginal neighborhoods in cities), makes it possible to assess each State's acts of commission and omission. However, the failure to comply may result in the Federal Executive Branch, or even the Legislative Branch, becoming permanent violators of human rights for not acting in accordance with law and opposing the norms of an overarching hierarchy (Gutierrez 2007).

It is likely that the UN Council of Human Rights will continue to emphasize the responsibility of States to assure full realization of human rights, while insisting that this achievement must be gradual and accompanied by international aid and cooperation in order to achieve the total realization of the right to drinking water and sanitation. Through these means, States must adapt their legislation to make their commitment in this matter explicit. In addition, it points to the important role of States in planning actions conceived as instruments for promoting human rights, including the right to drinking water and sanitation.

In sum, the Council calls on States to make steady progress towards achieving this right through concrete actions and measures in their countries. Undoubtedly, this is a fundamental starting point for evaluating compliance and the General Assembly calls on States to continuously monitor and analyse progress, assess existing policies and programmes, develop comprehensive plans and strategies, bring existing legislation in line with the human right, to ensure transparency and

[3] See Box 4.1.

the maximum available resources. Furthermore, States ought to provide a regulatory framework which protects and respects the human right, and to provide legal remedies and access to justice (UN General Assembly 2011: 3–4). These policy goals have been echoed in reports by the Special Rapporteur which also contains specific recommendations for States with regard to their obligation to integrate changes in their public policies and planning and financial strategies to make this right a reality (UN General Assembly 2013: 22) as well as a typology of violations (UN General Assembly 2014: 23). Certain recommendations in this report seek to ensure that violations are not only identified, but can be prevented or remedied with the intention to prevent continuous or systematic violations by States and enforce compliance with this human right in their countries through international monitoring and citizen participation (UN General Assembly 2014: 23–25).

Most recently, the UN General Assembly accepted new commitments that have clear implications in relation to public policies for the next 15 years, and are stipulated in the Post-2015 Development Agenda. One of this Agenda's 17 objectives is directly related to the realization of the human right to water, as it proposes ensuring availability and sustainable management of water and sanitation for all people (UN General Assembly 2015: 18–19). It calls on the global community to:

- *"By 2030, achieve universal and equitable access to safe and affordable drinking water for all.*
- *By 2030, achieve access to adequate and equitable sanitation and hygiene for all and end open defecation, paying special attention to the needs of women and girls and those in vulnerable situations.*
- *By 2030, improve water quality by reducing pollution, eliminating dumping and minimizing release of hazardous chemicals and materials, halving the proportion of untreated wastewater and substantially increasing recycling and safe reuse globally.*
- *By 2030, substantially increase water-use efficiency across all sectors and ensure sustainable withdrawals and supply of freshwater to address water scarcity and substantially reduce the number of people suffering from water scarcity.*
- *By 2030, implement integrated water resources management at all levels, including through transboundary cooperation as appropriate.*
- *By 2020, protect and restore water-related ecosystems, including mountains, forests, wetlands, rivers, aquifers and lakes.*
- *By 2030, expand international cooperation and capacity-building support to developing countries in water- and sanitation-related activities and programmes, including water harvesting, desalination, water efficiency, wastewater treatment, recycling and reuse technologies.*
- *Support and strengthen the participation of local communities in improving water and sanitation management."*

It is hoped that with the UN's recommendations for the countries that have ratified the resolution on the human right to water, and the commitments assumed through the Post-2015 Development Agenda for the realization of this right, that

States will proceed to modify and align their legislation and public policies in the coming years. In this respect, several countries in the Global North and South have explicitly and formally recognized this basic human right in their constitutions and national laws. Moreover, they have enriched jurisprudence by processing concrete cases leading to resolutions in national and international tribunals.

For example, various Latin American nations – Uruguay, Ecuador, Bolivia, Nicaragua, Honduras, the Dominican Republic and Mexico – have included the human right to water in their constitutions, while others (eg. Paraguay, Honduras, Nicaragua) have enacted laws related to water that explicitly include this right. At the level of the application of jurisprudence, countries like Argentina, Costa Rica, Colombia and Mexico have shown advances in some litigation on violations of the human right to water. It is important to point out that Brazil is the only country in this region that has refused to recognize the human right to water, even abstaining from voting on the Resolution of the UN General Assembly in 2010 (Mora and Dubois 2015).

4.4.2 *Civil Society and the Human Right to Water: Some Examples of Violations and Achievements in Latin America*

Changes in the laws are not sufficient if public policies lack a focus on human rights, or if national plans and programs omit full realization of the human right to water, which entails greater funding and concrete actions that benefit the most vulnerable population sectors. But the same can be said if citizens do not have guaranteed access to the tribunals of justice when violations of their human rights occur, or if they are impeded from participating in decision-making and the construction of public policies or laws.

The situation regarding non-compliance of the human right to water in Latin America is reflected in numerous conflicts over this resource in which affected populations have been forced to recur to both legal defense measures and social mobilizations to demand respect for their essential human rights. Here, the role of States has been contradictory: on the one hand, together with the private sector they foment private projects and financial investments that affect people's human rights (for example, open-pit mining, fracking technique to extract natural gas, large hydroelectric dams); while on the other they sign and ratify international resolutions regarding human rights and modify their constitutions and national legislation.

Unfortunately, in the past decade States have restricted access to justice and respect for law to such an extent that affected populations have been forced to take their cases to international instances, either formal (like the UN), or informal (eg. ethics tribunals). The most critical cases in Latin America have occurred in Mexico, Central America, Brazil, Peru and Chile due to the expansion of open-pit mining by U.S. and Canadian companies that has contaminated springs, rivers, lakes and seas, and affected the lands of indigenous and peasant communities.

Indeed, the establishment of the Latin American Water Tribunal is explained by the increase in conflicts over water in the region and the denial of access to environmental justice. From 2000 to 2016, this Tribunal has held 8 international audiences that have documented almost one hundred cases of such conflicts in Latin America. The most alarming situation is in Mexico, where systematic violations of human rights associated with water and the criminalization of social protests have been identified. In 2012, the National Assembly of Environmental Victims (*Asamblea Nacional de Afectados Ambiental*) presented the Mexican case before that Tribunal, signaling federal government agencies (Ministry of the Environment and the National Water Commission, among others) as those responsible for that country's water crisis. In its non-binding verdict, the Tribunal's jury found the following: the incapacity of the State (executive, legislative and judicial authorities) to establish effective mechanisms of access to environmental justice; and a growing deterioration of the human right to water in the country and the insufficiency of its juridical-political mechanisms to stop it. The Tribunal issued an alert because of the levels of social conflict throughout the country and the absence of instruments of citizen participation to channel it (TLA 2012).

In addition to handing down this decision, the Tribunal recommended that the Mexican State should respect laws, procedures, public policies and daily practices related to access to water and the environment to guarantee this as a fundamental human right and as a shared social good. It also emphasized the need to adapt them to the norms of equity, efficiency and sustainability as demanded by the country's Constitution and the International Treaties it has signed and ratified. It further recommended a review of the system for the procuration of environmental justice and the establishment of jurisdictional organs with full administrative and budget autonomy that would be in charge of tutelage and the application of environmental laws in accordance with the principle of the separation of powers sanctioned by the Constitution (TLA 2012).

It is important to mention that the Tribunal based this verdict on international jurisprudence and the universal recognition of the Human Right to Water and the UN Resolution that recognizes this right. It also referred to earlier verdicts formulated at previous Audiences, like those of 2006 and 2007, where many cases from Mexico were heard (TLA 2012).

Finally, at the level of the United Nations, in 2015 the international organization *France Libertes* presented – in writing – the case of the violation of the human right to water and sanitation in Latin America that resulted from the policies governing extractive activities (UN Human Right Council 2015). That text describes the critical situation in various countries, like Mexico, Guatemala and Honduras, due to their policies that regulate the extraction of minerals and natural gas, activities that affect water resources and the human rights of nearby populations (see Box 4.2). With international law on its side, that group issued a series of recommendations to those States which demand that they comply with, and enforce, the human right to water. It also called on the UN's Special Rapporteurs for the human right to water and indigenous rights to pressure States to respect these rights. It ends its presentation with suggestions as to how to advance in these areas (eg. organizing groups to work with transnational corporations and private companies).

Box 4.2 Violation of the Human Right to Water and Sanitation in Latin America: A Consequence of Policies Extractive Activities

"With the growing scarcity and depletion of natural resources, businesses have to diversify their zones of exploitation but also their techniques. In this struggle for raw materials, Latin American countries are not an exception and they bear the cost of extreme and expensive methods, also known as non-conventional, such as fracking. In December 2013, Mexico decided to put an end to seventy-five years of state energy monopoly on natural resource exploitation to open the sector to new extractive projects based on non-conventional methods. This leads inevitably to intensive water use, with serious consequences for local communities' fundamental right to water: contaminated surface and ground water rendering it non-potable, increased water stress, competition between different water uses, serious health issues such as poisoning, skin diseases and malformations.

In Mexico (in Sierra Norte de Puebla region), in Guatemala (Petén region) and in Honduras, concessions to extractive industries have been awarded in extensive areas. In these countries, local communities have difficulties in expressing their concerns. They are either completely absent in decision making processes or only involved in late stages, the dialogue remaining between the State and businesses.

In Latin America, many civil society organizations get together to mobilise around the harm done to their environment and to defend their fundamental rights in States where governments is in collusion with businesses the extractive industries. We denounce the devastating effects of extractivism, which are worsened by an increasing criminalization of social movements protesting about large-scale industrial projects."

Source: UN Human Right Council (2015)

4.5 Final Comments

From an integrated perspective, the human right to water is defined as the responsibility of society to guarantee the satisfaction of basic needs for both present and future generations, without affecting ecosystems and hydrologic systems. This implies that State regulation is necessary to reach the goal, together with the participation of private and social stakeholders and actors. This vision seeks to attain international cooperation and solidarity in order to satisfy these needs in all countries, regardless of their economic, political or religious conditions. It also considers environmental aspects by recognizing that satisfying those needs entails protecting the natural base that supports life on the planet.

It is important to emphasize that satisfying basic needs goes beyond guaranteeing an adequate water supply, for the water available must also be of sufficient quality, and access must be open to all people. Likewise, it includes food production for subsistence, the appropriate use of irrigation water, and the conservation of aquatic ecosystems (lakes, mangrove forests and rivers where fishing is practiced), democratic water management through social participation in decision-making regarding uses and distribution, and the conservation of water resources and ecosystems. It further entails guaranteeing respect for development modalities, lifeways, and community identities in territories with peasant and indigenous populations where the sustainable use and management of ecosystems and water resources depends on uses and customs that preceded state legislation.

While the integrated perspective on water as a human right is an ideal to be reached by society, in practice it has been simplified to satisfying the basic need for clean drinking water. In other words, the complexity and interconnection of water with development, social wellbeing, health and environmental protection, cultural identity, political participation, and self-determination, has been narrowed to just guaranteeing an adequate water supply to satisfy the essential needs of the population. This simplification allows States to assume their commitment to respect the right to water by creating, reforming or improving laws –political and institutional– and implementing mechanisms of compensation and governmental responsibility to guarantee water to the entire population.

Problems arise when States do not comply with international agreements through acts of omission or commission, and thus become permanent violators of human rights; a situation that currently characterizes most Latin American States (especially Mexico) and is reflected in the diversity and frequency of conflicts over water in this region.

The perspective that stresses the importance of complying with international agreements and resolutions concerning the human right to water and sanitation, and of achieving the Post-2015 Sustainable Development Agenda (eg. ensuring the availability and sustainable management of water and sanitation for all), holds that it is essential for States to assign larger budgets to solve the problems of poverty and social marginality (UN Water 2014). By doing so, advances would be made towards forging respect for such essential and elementary rights as access to water for all people, and States would no longer be permanent violators of international agreements and commitments.

Through social pressure and political lobbying it is possible to reorient the acts of the State towards implementing legal measures and public policies that contribute to resolving the essential problems of the population, such as supplies of drinking water. In this regard, the role of civil society and social movements is essential in remodeling the priorities of States towards defending the collective interest and respecting human rights, and it is in that context that ethical tribunals gain relevance as alternative instances of environmental justice and defense of the human right to water.

References

Avila, P. (2007). *El manejo del agua en territorios indígenas en México, Serie del Agua en México, vol. 4*. Mexico: Departamento de México y ColLatinoamérica y el Caribe, Banco Mundial.

Bakker, K. (2007). The commons versus the commodity: Alter-globalization, anti-privatization and the human right to water in the global south. *Antipode, 39*(3), 430–455.

Bogantes, J. (2007). Tribunales éticos: un acercamiento filosófico y práctico a la justicia ambiental. In S. Esch et al. (Eds.), *La gota de la vida: hacia una gestión democrática y sustentable del agua* (pp. 150–157). Mexico: Fundacion Boll.

Chociej, Z., & Adeel, Z. (2012). Legal and ethical dimensions of a right to water. In H. Bigas (Ed.), *The global water crisis: Addressing an urgent security issue* (pp. 122–127). Hamilton: United Nations University Institute for Water, Environment and Health (Inter-Action Council).

Gleick, P. (1999). The human right to water. *Water Policy, 1*(5), 487–503.

Gutiérrez, R. (2007). El derecho humano al agua en México. In S. Esch et al. (Eds.), *La gota de la vida: hacia una gestión democrática y sustentable del agua* (pp. 71–90). México: Fundacion Boll.

Hernandez, R., & Herrerías, G. (2005, 11–14 July). "Water Forever Program: Executive summary", *5° Simposio Brasileño de captación y manejo de agua de lluvia,* Teresina, Brasil: Piauí

Jennings, B., Heltne, P., & Kintzeke, K. (2009). Principles of water ethics. *Minding Nature, 2*(2), 25–28.

Langford, M., & Khalfan, A. (2007). Introducción al agua como derecho humano. In S. Esch et al. (Eds.), *La gota de la vida: hacia una gestión democrática y sustentable del agua* (pp. 30–62). Mexico: Fundacion Boll.

Leb, C. (2012). The right to water in a transboundary context: Emergence of seminal trends. *Water International, 37*(6), 640–653 http://www.tandfonline.com/doi/abs/10.1080/02508060.2012.

Mora, J., & Dubois, V. (2015). *Implementación del derecho humano al agua en América Latina.* Seoul: Corporación Andina de Fomento. Seventh World Water Forum.

Murillo, D. (2014). Las dimensiones eludidas en el derecho humano al agua y al saneamiento actual en México. In R. López, J. L. Martínez, & E. López (Eds.), *Viabilidad y Barreras para el ejercicio del derecho humano al agua y saneamiento en México* (pp. 37–73). México: Instituto Mexicano de Tecnología del Agua.

Murthy, S. (2013). The human right(s) to water and sanitation: History, meaning, and the controversy over privatization. *Berkeley Journal of International Law, 31*(1), 89–147.

Pardy, B. (2011). The dark irony of international water rights. *Pace Environmental Law Review, 38*(3), 907–919.

Pardy, B. (2012). False Panacea: The human right to water. In H. Bigas (Ed.), *The global water crisis: Addressing an urgent security issue* (pp. 136–141). Hamilton: United Nations University Institute for Water, Environment and Health (Inter-Action Council).

Plurinational State of Bolivia. (2010, 21 December). *Law of the Rights of Mother Earth (Law 071),* Plurinational Legislative Assembly.

Sandoval, A. (2001). *Los derechos económicos sociales y culturales. Una revisión del contenido esencial de cada derecho y de las obligaciones del Estado.* Asociación Latinoamericana de Organizaciones de Promoción, México: Equipo Pueblo A.C.

Scanlon, J., Cassar, A., & Nemes, N. (2004). *Water as a human right?* Gland/Cambridge: IUCN.

Second World Water Forum. (2000). *Ministerial declaration of The Hague on water security in the 21st century.* The Hague: Springer.

TLA. (2012). *"Caso: Posible colapso hídrico en México y violación del derecho humano al agua de sus presentes y futuras generaciones, República de los Estados Unidos Mexicanos". Audiencia Tribunal Latinoamericano del Agua.* Argentina: Buenos Aires.

UN Development Program. (2006). *Human development report 2006. Beyond scarcity: power, poverty and the global water crisis.* New York: UNEP.

UN Economic and Social Council. (2002). *General Comment No. 15: The Right to Water. Articles 11 and 12 of the International Covenant on Economic, Social and Cultural Rights*. United Nations- Committee on Economic, Social and Cultural Rights, Geneva, November, E/C.12/2002/11.

UN General Assembly. (2010). *The human right to water and sanitation: resolution adopted by the General Assembly*, United Nations, July, A/RES/64/292.

UN General Assembly. (2011). *Resolution adopted by the Human Rights Council: 18/1 The human right to safe drinking water and sanitation*, United Nations General Assembly- Human Rights Council, October, A/HRC/RES/18/1.

UN General Assembly. (2013). *Report of the Special Rapporteur on the human right to safe drinking water and sanitation, Catarina de Albuquerque*, United Nations- Human Rights Council, July, A/HRC/24/44.

UN General Assembly. (2014). *Report of the Special Rapporteur on the human right to safe drinking water and sanitation, Catarina de Albuquerque*, United Nations- Human Rights Council, June, A/HRC/27/55.

UN General Assembly. (2015). *General outcome document of the United Nations summit for the adoption of the post-2015 development agenda*. August. A/69/L.85.

UN Human Rights Council. (2015). *Violation of the human right to water and sanitation in Latin America: a consequence of policies extractive activities*, Joint written statement* submitted by France Libertes: Fondation Danielle Mitterrand, American Association of Jurists, Emmaus International Association, Peace Brigades International Switzerland, non-governmental organizations in special consultative status. February. A/HRC/28/NGO/60.

UN Water. (2012). *Report 4: Managing water under uncertainty and risk: Executive summary*. Paris: United Nations World Water Development- UNESCO.

UN Water. (2013). *Water security & the global water agenda: A UN-water analytical brief*. Ontario: United Nations University- Institute for Water, Environment & Health.

UN Water. (2014). *A post-2015 global goal for water: Synthesis of key findings and recommendations from UN-Water*. United Nations: Water.

UN World Health Organization. (2003). *Right to water*. France: World Health Organization.

UN World Health Organization. (2010). *Right to water* (Fact Sheet No. 35). Geneva: United Nations Human Rights-UN Habitat-World Health Organization.

Chapter 5
Crime, Corruption, Terrorism and Beyond: A Typology of Water Crime

Kyungmee Kim and Ashok Swain

Abstract Water security is a multifaceted concept that spreads over political, social, economic and biophysical fields, becoming increasingly important in the age of looming global water crises. Previous research outcomes suggest that escalating global water crises are the results of governance failure not limited to physical shortages of freshwater resources. Water crime is defined in both procedural and moral terms as wrongdoings determined within the legal justice systems and social norms. This chapter explores water crimes in different dimensions with examples from the Global North and South and establishes typologies as follows: mismanagement of water resources causing significant social harms and environmental damage; corruption allowing allocation of water resources for a favoured party, using public office for private economic and political gains and adding payments for more effective service delivery; and, terrorism targeting water infrastructure and systems and affecting water security in water scarce regions. A broader framework to understand the multiple dimensions of water crime is an essential precondition for establishing a comprehensive strategy for achieving water security.

Keywords Water • Security • Crime • Governance • Management • Terrorism

5.1 Introduction

Freshwater constitutes only 3% of all water on the planet, and only one-third of it is available for use as the rest is tucked away in glaciers and snowfields. Freshwater is critical to human survival and wellbeing, but this vital resource is unevenly distributed. Many of the water systems supporting an increasing human

K. Kim (✉) • A. Swain
Department of Peace and Conflict Research, Uppsala University, Uppsala, Sweden

Research School for International Water Cooperation, UNESCO Category II Centre on International Water Cooperation, Uppsala, Sweden
e-mail: kyungmee.kim@pcr.uu.se

D. Devlaeminck et al. (eds.), *The Human Face of Water Security*, Water Security in a New World, DOI 10.1007/978-3-319-50161-1_5

population and the ecosystem have been under stress from overexploitation (Swain 2004). Access to drinking water is considered as a universal right to all humanity, but approximately one in nine people do not have access to safe water and 2.5 billion people lack access to improved sanitation, causing the death of 842,000 people each year, mostly children from diarrheal diseases (Prüss-Ustün et al. 2014). Climate change has further escalated this perilous situation by altering patterns of weather, water availability and demand around the world (Swain 2012). In this context, water crises are being viewed as the most serious risk to business and society by the Global Risk 2015 Report prepared for the World Economic Forum.

Scientific research outcomes started during 1990s have addressed that increasing water scarcities have affected water management on all levels of society, and that pressure on water resources for human needs and ecosystem services has been incremental (Falkenmark 1990; Falkenmark et al. 1990). Water is critical for producing food and the source of livelihood for 70% of the population from developing countries. Freshwater serves a vital role in the ecosystems such as in rivers, lakes and wetlands. Milly et al. (2008) argue that the stationarity of water, the key assumption that natural systems fluctuate within an unchanging envelope of variability, has been dismissed due to the anthropogenic activities in river basins and climate change. The human activities and modification of watersheds by dams, reservoirs and diversions have altered the hydrological cycle on a significant scale. Coupled with population growth, available freshwater per capita is drastically decreasing, on the other hand, demand for water is rapidly increasing for industrial purposes. Degradation of freshwater ecosystems reinforce the cycle of decline and disturb ecosystem services.

Water is a complex resource with unique characteristics of limited transferability, seasonal and intra-annual variability, and irreplaceability. A water crisis is not only physical water shortage, but also a lack of access to clean water caused by economic, social and political constraints. In addition to the biophysical aspect, some focus on the supply and demand side while policy and management fail to properly respond to the impending crises (Biswas 1997; Lundqvist 2000). International policy frameworks have been developed to address the global water crises. The United Nations has been raising public awareness on looming global water scarcity since the United Nations Conference on the Human Environment in 1972. The discussions on water issues have continued through the United Nations Conference on Environment and Development in Rio de Janeiro (1992) and the United Nations Conference on Sustainable Development (2012).

Water security has emerged within discussions on ‘alternative security’ attempting to broaden the security agenda (Buzan 1991; Klare 2001; Jägerskog et al. 2014). Prior to the 1990s, the discussions on security mainly took place in strategic studies focusing on aspects of military and defence. The broadened security agenda during the beginning of the post-Cold War era was more inclusive to the security threats coming from political, social, economic, and environment sector. Buzan (1991) finds security refers to “the ability of states and societies to maintain their independent identity and their functional integrity” and notes that the environmental changes will transform the security environment in all sectors, perhaps even at all levels of society. Security is often defined by threats and vulnerabilities. Water is fundamental to human survival and development and an essential component to well-functioning ecosystems. As such, a severe water shortage can be a threat to the functional integrity of vulnerable societies. Connecting water and security concepts has been done by policy actors and academics. For example, the Global Water Partnership (2010) defined water security as an overarching goal where "every person has access to enough safe water at affordable cost to lead a clean, healthy and productive life, while ensuring the environment is protected and enhanced."

Water security has four broad aspects: first, a constant availability and access to water should be guaranteed; second, avoiding excessive quantities of water is essential to prevent water related natural disasters; third, an acceptable quality is required for water use; and fourth, satisfying human needs is the basis for achieving water security. The definition by Grey and Sadoff (2007) embraces the first three aspects, that water security is "the availability of an acceptable quantity and quality of water for health, livelihoods, ecosystems and production, coupled with an acceptable level of water-related risks to people, environments and economies." Witter and Whiteford (1999), on the other hand, extend the definition to include human needs "necessary, at an affordable price, to meet both the short-term and long-term needs to protect the health, safety, welfare and productive capacity of position." In a review of the water security concept, Cook and Bakker (2012) point out that none of the existing terms capture the complexity of water-related issues and that an integrative framing can be beneficial to include various stakeholders and human and ecosystem health concerns. As defined above, water security conceptually penetrates the multiple sectors and users. The concept plays an important role since the conceptual boundary of water security influences policy goals and interventions.

The water security problem is a complex multilevel policy goal that requires the efforts of all levels in society, even those beyond the water sector. Governance is one of fundamental challenges where both technical and ethical questions are raised. Mismanagement of water is never a purely technical problem but a consequence of governance failures such as negligence, corruption and insufficient institutional and human capacity. Mismanagement has dire consequences on human well-being and ecosystems. Recognising governance failures as a key threat to water security opens the discussion on the criminal aspect of water security. Water crime as a term explores the connection between ethical norms and legal frameworks, emphasizing questions on accountability in water security issues.

This chapter will provide an overview on water crime, a term which lacks a clear definition. Previous research on the criminal aspects of water have focused on legal frameworks and compliance mechanisms (Franz 2011; Burns et al. 2013). Laws and regulations are crucial in the operational aspect of water crime. Compared to other natural resources, water is neither easily exploitable nor transferrable due to its sheer volume and weight, and is irreplaceable for human consumption, food production and industrial activities. Unlike other natural resources these aspects force us to apply the conventional frameworks of crime, corruption and terrorism in the water sector. Water crime can have drastic impacts beyond water on food and energy security, livelihoods and the environment.

The earliest evidence of criminalising misdeeds in water management can be found in the Code of Hammurabi created in 1785 B.C.E (King 1910). The Code states that one shall be punished for negligence in managing its own irrigation canals that caused damage to adjacent fields. The primary modern definitions of crime are in two categories, the moral and procedural (Farmer 2008). Moral definitions focus on moral wrongdoings in criminal activities, or *mala in se*. On the other hand, procedural definitions define crimes as "an act capable of being followed by criminal proceedings having a criminal outcome, and a proceeding or its outcome is criminal if it has certain characteristics which mark it as criminal (Williams 1955)," or *mala prohibita*. Traditional ideas of moral consensus and moral relativism have influenced moral definitions of crime, however legal justice, social harms and the scope of criminal laws have influenced the development of procedural definitions.

Environmental crime can be defined in procedural, legal and moral terms. For example, the European Commission defines environmental crimes as "acts that breach environmental legislation and cause significant harm or risk to the environment and human health" (European Commission 2015). This definition emphasises the criminal acts as violating procedural norms while causing social harms. On the other hand, public and media scrutiny on environmental crimes is not only limited to the violation of laws, but also focuses on the impact on environment and social harms. Increased public awareness on the role of the private sector for environmental and social responsibility plays an important role in extending public scrutiny to the broader scope of environmental crimes. Defining water crime, in this context, encompasses the moral and procedural definitions of crime. For example, industrial operations violating laws and regulations are water crimes regardless of the level of consequential social harms. The public, however, can recognise some activities that cause significant damage to the environment as 'crimes' although these activities do not violate the law.

Corruption is one of the major causes of the water governance crises contributing negatively for achieving water security. In the water sector corruption is widespread globally throughout water resource management, drinking water and sanitation, irrigation and the hydropower sector (Transparency International 2008). Most transactions driven by corruption are unlawful and even criminal since they violate rights and cause significant social harms. Corruption in water and its related sectors occur in various forms, scopes and scales. Defining corruption is problematic because its definition depends on what is modelled and measured (Jain 2001). The most widely used definition of corruption is "the abuse of entrusted power for private gain" by Transparency International. Corruption influences a wide range of government policies on economic spending on water related infrastructure, water rights and allocations and management practices of water systems. In this chapter, water corruption will be categorised into three different forms: political, economic and bureaucratic corruption.

'Water terrorism' has become one of the emerging concerns in regions experiencing severe water scarcity and escalating armed conflicts. The link between water and terrorism can be investigated on at least two dimensions. First, a deliberate act to destroy or disrupt water infrastructure may be directly responsible for damaging areas of 'enemy' groups or inducing water scarcity for the opponent. Terrorist groups might also poison or pollute water storage and distribution facilities in an attempt to cause harm to the opponent. Thus, water becomes one of the strategies by terrorist organisations. Second, water has broader linkages to terrorism. Growing water scarcity threatens the livelihoods and human security in arid regions and increases the potential to facilitate a socio-economic setting for terrorism to germinate and sustain itself.

Global water crises cause greater water scarcity and undermine the effort to achieve access to clean water for all. Water consumption is rising steeply, and freshwater sources are becoming depleted and unreliable for human consumption in many parts of the world. Water tables are falling on every continent while rainfalls and extreme weather events are less predictable. Water scarcity and extreme weather events may force states, riparian states, communities and individuals to take conflicting positions. In such situations of power asymmetry, one party may try to engage in terrorist acts or support terrorist groups as part of its strategic asset, vis-à-vis its powerful enemy. Terrorism over water can be observed at different levels of society. Water issues can create terror groups within a state and between riparian states as water disputes tend to become more complex and difficult when international water systems are shared with a hydrohegemon.

5.2 Water Crime by Mismanagement of Water

Criminal aspects of mismanaging water are closely associated with the consequences causing significant social harms. Environmental laws and regulations aim to control and monitor the impact of human activities on the environment. Depleting freshwater water resources and declining water quality are critical reasons for law and regulation to manage water resources. Water laws are tools in regulating water allocations, water quality, water quantity, and management of water sources including protection and conservation of watersheds. In order to ensure the quality of water, water laws comprehensively dictate the regulations on the discharge of domestic sewage, animal and agricultural wastewater and hazardous substances from industrial activities to watersheds. Compliance, monitoring and assistance mechanisms include on-site inspections, evaluations and investigations, and off-site data collection, review, reporting, programme coordination, oversight, and support.

There are many forms of disobedience of compliance regimes, here three types are introduced as follows: (1) breach of a water licence, water use permit, or allotment by individuals, households or larger business establishments so called 'water theft'; (2) insufficient wastewater treatment systems and illegal disposal of pollutants and wastewater into sewer systems and watersheds by households and business establishments; and (3) falsification in reporting on water quality. Water thefts often occur in locations experiencing regular water shortages in both developing and developed countries. According to media reports, water thefts have become increasingly common in the US state of California as the state is experiencing its worst drought on record, so much so that the state authority declared a drought State of Emergency in January 2015 (US State of California 2015). One of the cases in California involves individuals stealing water from fire hydrants (see Baxte 2014; Bender 2015; and Foran 2014).

Water theft also includes tampering with water taps in order to bypass water meters and disconnecting or connecting from the public water supply systems without permits. In agricultural water use, water theft is a common and reoccurring issue that affects the water use efficiency in canal irrigation systems (Gelles 2000; Gorantiwar and Smout 2005; Ray 2011). VanderMeer (1971) outlined seven ways of water thievery in canal irrigation as well as the frequency and factors affecting the behaviour. When irrigation water supply is the critical factor to determine crop yields, farmers are keen to maintain the irrigation water supply to their fields. Irrigation canals provide water supply to the fields that are distant from direct water sources, yet the distribution and delivery of water to the field varies by the location of farmland. Upstream farmers can obstruct the waterway to prevent the flow of water downstream, or some farmers may have priority access to irrigation water due to their influence.

Falsifying water related data and reporting is a criminal activity under law. Pollution and quality control of water is heavily dependent on the system to review and monitor reports and data from industry or regulatory bodies. Various types of falsification of data and statements with false information can be subjected by law enforcement if the mismanagement had a criminal intent and caused harmful social impacts.

Water laws and regulations are important tools that provide authorities to operationalize the goals to ensure water security for all, but limitations exist in adapting to new threats and challenges as well as broadening the policy to integrate other environmental issues connected to water. Innovations and new technologies have provided new ways of developing natural resources. However, these new developments sometimes have harmful impacts on water resources and the environment. For example, laws have permitted mining and hydraulic fracturing (or fracking) for natural gas extraction in many countries. Hydraulic fracturing is a technology to extract natural gas stored in shale by injecting large amounts of water with additives into the well drilled in the shale layer. This acts to pressurise networks of fractures in order to bring the natural gas to the earth's surface. On average, 20 million litres of water and 200,000 l of acids, biocides, scale inhibitors, friction reducers and surfactants are injected in each well (Howarth et al. 2011). The significant volume of water used for fracking imposes the risk of ground water contamination. As such, some policy makers call for the ban of fracking methods at the state level in the United States. For instance, Holzman (2011) found the evidence of methane contamination in private wells near fracking sites. Aldhous (2012) points out the comprehensive risk on human health caused by fracking through air pollution, groundwater contamination, and illegal discharge of pollutants, and accidents. The Government of France enacted a ban on fracking in 2011 and its constitutional court rejected an appeal to overturn the ban in 2013 (BBC 2013). Cases of moratorium or legal ban on fracking can be found at the subnational level in Australia, United Kingdom, and United States (see ABC News 2015; BBC 2015; The Guardian 2015). Discussions on fracking present challenges in developing water resources while balancing the needs for energy security, the protection of watersheds and ecosystems and, safeguarding clean water for human health. The water footprint of the energy production is significant overall, but the lack of data makes it difficult to quantify the trade-offs between water and energy production from other energy sources.

5.3 Water Corruption as Water Crime

A number of scholars have suggested typologies of corruption through an empirical analysis (e.g. Roebuck and Barker 1974),[1] degrees of moral violation (Heidenheimer 1989), scales and methods (Punch 2000). In this section, water corruption is categorised into grand and petty corruption. Grand corruption refers to corruption occurring in a larger scale often with political back-ups or corruption activities which influence high-level policies and decision-making processes. Grand corruption is committed by or involves elites who would exploit their power to make economic or political gains. Distinctions between active and passive corruption are more important in a legalistic sense while the corruption is conducted with deliberate motives and decision-making processes. Petty corruption refers to an exchange of small amounts of money or valuable commodities for minor favours. Petty corruption appears to be common in developing countries and has a bureaucratic nature. Mid-level to low-level public administration is often involved with petty corruption rather than high-level.

5.3.1 Grand Corruption

Political corruption pervades in high-level public offices and political elites who are motivated to achieve their political aims and agenda through corruption. Peter and Welch (1978) assume that "political behaviour is corrupt when it violates some formal standard or rule of behaviour set down by a political system for its public officials." Their definition of political corruption limits the scope of corruption within illegal activities, but a broader definition of political corruption includes not

[1] This author delineates eight types of crimes by police including corruption of authority, kickbacks, opportunistic theft, shakedowns, protection of illegal activities, underhand arrangements, direct criminal activities and internal payoffs. See: Roebuck, J. B., & Barker, T. (1974). A typology of police corruption. *Social Problems*, *21*(3), 423–437.

only procedural crime but also compromises of morality and integrity. Vargas-Hernández (2009) defines economic corruption as "the sacrifice of the principal's interest for the agent's interest." Economic corruption refers to the acts of transferring public wealth for private gains. Economic incentives are strong motivations in corruption cases when the lack of anti-corruption measures prevail. Public administration or bureaucratic corruption refers to the behaviours of public agents neglecting the principles of efficiency, truthfulness and rightfulness. Public administration corruption causes the transfer of public benefits to private gains by taking advantage of the entrusted power, as for example, in the form of nepotism that results in the transfer of benefits from society to family members (Vargas-Hernández 2009).

Water as a vital natural resource is an important instrument to advance political aims and agendas. Water related infrastructure such as hydropower dams, irrigation systems and public water utility systems can be a target or driver of political corruption. High-level politicians and officials can influence the policy level of public spending, including infrastructure development where uneven political and economic power can immensely influence the outcome of decision-making (Johnston 1997). Privatisation of public services including water utility systems, hydropower sector and irrigation systems have become a trend in recent decades. For example, infrastructure development has increasingly adapted Build-Operate-Transfer (BOT), Build-Own-Operate-Transfer (BOOT), or Build-Lease-Transfer (BLT) schemes for private developers and operators in public investment projects. Procurements for public infrastructure involve significant amounts of financial incentives for the private sector where public agents can be key actors and may be politically and economically motivated.

One of the relevant cases of water related corruption involving high-level decision makers is the phenomenon of water grabbing through large-scale land deals. The food price crisis of 2007–2008 and 2011–2012 is one of the drivers of growing interest in farmland as an economic asset in Africa, Asia and Latin America (Cotula 2012). Land Matrix (2015) identified 1043 concluded land deals covering more than 38 million hectares globally. Without an adequate and sustainable water supply, agricultural projects are far from viable. Rulli et al. (2013) provided the quantified annual water requirements of the recent land acquisition claims (47 million ha) at approximately 31 billion cubic meters of green water (rainfed) and 14 billion cubic meters of blue water (water needed for irrigation). The blue water requirement accounts for approximately 10% of total freshwater withdrawal for the agricultural sector, which accounts for 271 billion cubic meters per year (FAO AQUASTAT 2010). Metha et al. (2012) argue that water is both a target and driver of recent land acquisitions and the impact of land deals influences the future use and benefits of water rights and changes in tenure relations.

The term 'land grabbing' involves normative value proposition assuming the process of 'acquisition' includes the violation of the rights of local populations. The term 'land acquisition' is value-neutral, while the term 'land investment' is associated as a positive value with financial transactions between investors and invested assets. 'Water grabbing' refers to situations when the rights to access water resources are taken away from communal users by a powerful actor using force or other unlawful measures without free, prior and informed consent of the affected users. 'Land grabbing' assumes that the decision making process behind the deals are corrupted by either political or economic powers. Corruption causes and affects land grabbing by creating governance failures at various levels (MacInnes 2012). Public officials are important agents allowing land grabbing by bypassing or ignoring safeguards or laws for the favoured parties.

Actors engaged in land grabbing are often governmental officials, politicians, private investment funds, multinational agricultural companies, domestic investors, consultants and other private sector actors providing services to the deals. Most large-scale land deals are signed between public offices, often government entities on the ministerial, provincial or district level, and investors of both foreign and domestic origin (Deininger and Byerlee 2011). In cases, where foreign investments are restricted, foreign investors collaborate with domestic companies to acquire licenses. Many African countries have adopted legal reforms to promote foreign direct investments including the legal safeguards against discriminatory and arbitrary treatment and the government facilitation for large-scale investments (Polack et al. 2013).

Sometimes the lack of transparency in land deals, e.g. non-disclosure of parties in contracts, limits the possibilities to inquire accountability to the responsible party. Polack et al. (2013) analysed the legal frameworks of 12 African countries in relation to promoting accountability in relation to large-scale land acquisitions and concluded that the legal shortcomings to protect the local farmers rights undermine the pathways to greater empowerment. One of the legal conditions constraining this is limited access to information. Only six African countries (Angola, Ethiopia, Liberia, South Africa, Uganda and Zimbabwe) have legislation to guarantee the freedom of information that would enable citizens to gain access to information held by government agencies (UNESCO 2012). These conditions lead to difficulties in establishing the accountability of governments and investors for the loss of land by local farmers and communities.

Another ambiguity found in large-scale land deals is the resources required for project implementation such as water and electricity. Cotula (2011) points out the lack of clear provisions in water supply in the contracts for large-scale land deals. In principle, foreign investors are subjected to the domestic laws for water use and land tenure rights, unless international investment agreements, including Bilateral Investment Treaties, provide additional rights (Smaller and Mann 2009). Measuring the impact of water grabbing on local livelihoods and environment is problematic if the provision and implementation of domestic laws are unclear and lagging behind.

Feasibility studies, Environmental Impact Assessment (EIA) and Social Impact Assessments (SIA) are often formal requirements in large-scale investment planning in many national legislatures. The scope and effectiveness of EIA and SIA can be largely limited because the outcomes and recommendations of these assessments are systematically not reflected in strategic decision-making process, i.e. earlier in the project planning period (Bina, 2007). A lack of monitoring components in EIA implementation is also problematic since by no means are there mechanisms to ensure monitoring of project impacts and to rectify impacts that were not previously identified though an EIA (Momtaz 2002).

At the policy level, political elites can promote foreign direct investments and other large-scale agricultural expansion schemes that justify large-scale land acquisitions. For example, state-driven biofuel policies in Mozambique have attracted foreign and domestic investors to grow water intense crops that have caused conflict due to competing land and water use by governments/investors and communities (Nhantumbo and Salomão 2010). Foreign Direct Investment (FDI) in agriculture has been increasing during last decade due to increased demand in agricultural commodities and investment interests by transnational corporations and state owned enterprises and changes in investment host-country policies on banning FDI in agriculture. Developing countries have lifted their ban on FDI in agriculture, and instead have begun to promote the influx of FDI to modernise and improve the agricultural sector (UNCTAD 2009).

Among other forms of political corruptions land and water grabbing has significant impact on the water and food security of the local population. Powerful actors can influence the policy making on land entitlement and FDI as well as the project approval processes including site selection, procurement and EIA that allows land acquisitions through large-scale agricultural projects. Challenges to govern land and water deals increase when there is a lack of transparency and information around contracts and projects. Overall, corruption is at the root of these governance challenges in various aspects.

Actors engaged in corruption aim to gain financially through inappropriate uses of power given to the offices entrusted by the public. Large-scale economic corruptions in the water sector involve a number of corruption methods such as bribery, collusion, embezzlement, theft, fraud, extortion, and abuses of discretion. Both private sector entities and public offices are involved in economic corruption in water and related sectors. Economic incentives are crucial elements in grand corruption cases, e.g. see the Lesotho Highland Water Project (Box 5.1).

Box 5.1 The Lesotho Highland Water Project

The Lesotho Highland Water Project (LHWP) is one example of a large-scale economic corruption case in the procurement of water infrastructure. LHWP is the largest water transfer scheme in Africa and supplies water to Johannesburg, South Africa from Lesotho while generating hydropower from the Orange-Senqu River. Phase 1 of the project created 4 dams and 110 km of tunnels with spending of two billion US dollars financed by the World Bank. In 1999, the civilian elected government in Lesotho commissioned an investigation and audit on the LHWP, after which it discovered a mega corruption case involving 12 multinational companies and a senior government official, the CEO of Lesotho Highland Development Agency. The multinational companies paid over two million US dollars to government officials for contract procurements (Haas et al. 2010).

5.3.2 *Petty Corruption*

Corruption in water and its related sectors occurs in various forms, scopes and scales. Institutional frameworks such as bureaucracy and public administration play an important role in corruption. Weber (1964) regards corruption as a State centered phenomenon reflecting the malfunctioning of a rationalized bureaucracy. He also assumes the model of bureaucracy that acts according to the rules and public interests in a democratic political system.

Petty corruption in public administration occurs when the public service delivery for water supply has low efficiency (Davis 2004). So-called "speed-up" money and occasional bribery for service activation is paid to public agents by consumers and users. Uncertainty and unreliability in water service delivery increases the willingness of consumers and users to cover small bribes as part of water service fees (Paul 1995). Payments to junior staff of public water and sanitation agencies by household members has also reportedly been used for expediting applications for new connections, quick attention to water supply and sewer repair work, and the provision or ignoring of illegal service connections. The water utility systems in Indian cities heavily rely on the human capacity for monitoring meters and service delivery, however field staff have limited financial incentives to maintain integrity while the implementation of legal punishment for anti-corruption is systematically insufficient.

5.4 Beyond Conventional Water Crime: Terrorism

Many point out that defining terrorism itself is problematic (Ruby 2002). Terrorism can be defined as both violation of ethics and legal procedures. Paukste (2004) concludes that terrorist offences are kinds of criminal offences with three characteristics: they involve violence, the targeting of innocent people, and are powered by ideological motivation. Water and natural resources motivate or influence terrorism in different contexts. The most established linkage between water and terrorism is water as a target of terrorism. Historically water has been a target or tool to achieve political and military aims (Gleick 2013). Gleick (2006) argues that water-related terrorism poses a real threat by possible attacks on water infrastructure and releases of chemical and biological agents into the water system. A framework linking terrorism to water security is useful to analyse the risks of terrorism to water resources. The framework provides three aspects: water as a target of terrorism, water as an encouraging factor of terrorism, and water as an instrument of terrorism.

During conflict, the deliberate targeting of water storage facilities by terrorist groups may be directly responsible for inducing water scarcity or reducing the water quality of the opponent. Thus, water scarcity becomes part of a terrorist group's strategy. Dams, water storage and conveyance systems can be targeted by terror groups. For example, in 1993 the Serbian militia intentionally damaged the Peruca

Dam in Croatia, causing a near collapse of the dam risking thousands of lives (Darnton 1993). In 2014, the Haditha Dam in Iraq was the target of ISIS terror, leading to air strikes by US forces. The Pentagon reasoned that "the potential loss of control of the dam or a catastrophic failure of the dam—and the flooding that might result—would have threatened U.S. personnel and facilities in and around Baghdad, as well as thousands of Iraqi citizens" (Collard 2014). This fear was not unfounded as ISIS had used its control of the smaller Fallujah Dam in early 2014 to flood bordering lands, and withhold water to south and central Iraq. It is not only the water storage and supply that can be affected by terrorist acts, fresh water scarcity also has the potential to cause or contribute to the emergence and/or escalation of terror activities by weaker states or ethnic groups. Water shortage can potentially threaten agricultural production and energy supply, adding to the challenges of weak and poor countries and segmented societies (Goldenberg 2014).

Global water demand is rising steeply, and the lack of adequate supply of water is a problem in many parts of the world. Water tables are falling increasingly on every continent. Lower water levels are increasingly evident in the Middle East, North Africa and South Asia which are all projected to experience water shortages over the coming years due to increasing population and inefficient water management. Several countries in these parts of the world already face serious problems in meeting rapidly increasing water demand. In this scarcity situation, shared water has increasingly become a source of tension. There are several examples of violent separatism and terrorist activities that have emerged over the issue of water. One clear case is the Cauvery River water dispute that resulted in violent acts by opposing groups in the southern part of India in late 1991, including several deaths and massive population displacement. South Asia is presently witnessing a number of violent internal conflicts over river water sharing. A dispute over the sharing of river water had also contributed to the violent separatist movement and terror activities in the Punjab province of India in the 1980s and 1990s. The Indus River water also plays a critical role in the Sind separatist movement in Pakistan (Swain 2009). The Middle East is also already suffering from chronic water shortages from decrepit water systems and deteriorating electricity supplies. But the conflicts and their aftermath, sanctions and security threats in the region escalated the risks in water resources and delivery systems that resulted in long-term health and humanitarian consequences (ICRC 2015). Many countries in the South are already suffering from a growing water shortage. Having weak state apparatuses and strong ethnic divisions, these countries might be further predisposed to similar water scarcity induced terror activities in the near future. As the statistics suggest, countries with a low supply of freshwater are more likely candidates for violent civil war than countries with a high supply of freshwater (Hague and Ellingsen 1998).

In spite of several decades of alarm-bell ringing, the global water crisis remains unresolved and constitutes a huge challenge to humankind. Global climate change has added increased uncertainties to the water supply and demand situation. With increasing temperatures, sizeable reduction in precipitation and melting of glacial sources of major freshwater systems, fewer water supplies will be available for the agricultural sector. Climate change will not only decrease the supply of freshwater, it may also enhance its demand in domestic, irrigation, industrial and ecological uses. It is true that the projected impacts of global climate change over fresh water supply might be huge and dramatic and may lead to civil war and state collapse. Already some research points to climate change induced water scarcity as a catalyst for the on-going violent civil war and brutal acts of terror in Syria (Kelley et al. 2015). With the high climate related risks for water resources and environmental challenges, water scarcity and injustice can be used as an instrument for terrorists to mobilize supporters and to recruit new members (Box 5.2).

Box 5.2 Water and Terrorism
Water scarcity/mismanagement causing terrorism

- Shortages, poor water quality, as well as lack of water storage causing drought and floods can increase regional instability and tensions that encourage acts of terrorism.

  Water scarcity/injustice used as an instrument for terrorism

- Chronic water shortages and unequal distributions can be used by armed groups to mobilize supporters to commit terrorist acts.

  Water resources as targets of terrorism

- Water infrastructure including dams, reservoirs and irrigation canals can be targeted by terrorist groups to strategically damage water systems and disrupt water supply to the population and, to a large extent, cause a mass displacement of population.

5.5 Conclusion

The concepts of crime, corruption and terrorism expand the criminal aspects in water security discussions. Intertwined with the global water crises, water security has become an important policy goal especially in countries with severe water scarcity. Achieving water security is a battle with many front lines including water crime and corruption. The scope of water crimes is not limited to the water sector. Water is a vital resource to agriculture, hydropower generation, industrial activities and domestic consumption. Widespread corruption in the water sector greatly affects many aspects of water security indiscriminately in developing and developed countries: grand corruption has significant impacts to water allocation, and therefore livelihoods and wellbeing of populations; and, petty corruption lowers public trust in water service delivery systems and hinders necessary water sector reforms and investments. Water crime as a consequence of negligence and mismanagement decreases the water quality, therefore reducing overall water availability. Legal compliance mechanisms are often challenging to implement in both the developed and developing world. Various risks on water by terrorism exacerbate the complex linkages between water scarcity and violent conflicts. Broadened definitions and extended conceptual frameworks linking criminal aspects of water security can benefit to integrate water crime, corruption and terrorism into water security. Further understanding water crime as a violation of law and regulation as well as a violation of ethical norms to cause significant social harms and environmental impacts enhances current legal and regulatory measures to protect and manage water resources.

Terrorism, on the other hand, is an emerging and significant threat to global security in recent decades. The linkages between terrorism and water are more intricate than the classical view which views water as a target of terrorist attacks. In addition, water can be a driver of terrorism as well as as an instrument to promote terrorism. In this context, crimes committed by terrorist organisations can affect the quality and quantity of water resources in areas where terrorism is prevalent. Humanitarian consequences of terrorism to water suggest that water is a crucial aspect of human security. Unlike other water crimes terrorism has a broader scope and impact that can span national borders or even cross-regions. Efforts to counter the risks of terrorism to water resources require more integrated and inclusive approaches with cooperation among actors in security, humanitarian, water resource management and the development sector. Enhancing the knowledge and understanding the linkages between water and terrorism alongside other water crimes can improve the overall approach to ensure water security in challenging situations.

References

ABC News. (2015). Fracking banned for five years by Tasmanian Government, *ABC News*, February 26, from http://www.abc.net.au/news/2015-02-26/fracking-banned-for-five-years--by-tasmanian-government/6265378. Accessed 5 Mar 2015.

Aldhous, P. (2012). Drilling into the unknown. *New Scientist, 213*(2849), 8–10.

Baxte, A. (2014). Water theft in California heightens state drought concerns. *PBS News Hour*, November 12, from http://www.pbs.org/newshour/rundown/concerns-california-drought-heightened-rise-water-theft/. Accessed 5 Mar 2015.

BBC. (2013). Fracking ban upheld by French court. *BBC*, October 11, from http://www.bbc.com/news/business-24489986. Accessed 3 Mar 2015.

BBC. (2015). Block on fracking in Scotland announced by minister. *BBC*, January 28, from http://www.bbc.com/news/uk-scotland-scotland-politics-31016537. Accessed 3 Mar 2015.

Bender, K. J. (2015). Water thefts on the rise in drought-stricken California. *Associate Press*, February 23, from http://abcnews.go.com/Technology/wireStory/water-thefts-rise-drought-stricken-california-29171843. Accessed 5 Mar 2015.

Bina, O. (2007). A critical review of the dominant lines of argumentation on the need for strategic environmental assessment. *Environmental Impact Assessment Review, 27*(7), 585–606.

Biswas, A. (1997). Water development and the environment. *International Journal of Water Resources Development, 13*(2), 141–168.

Burns, A., Gieringer, F., McAleer, D., & Mirabal, M. (2013). Environmental crimes. *American Criminal Law Review, 50*, 857.

Buzan, B. (1991). *People, state and fear*. New York: Harvester Wheatsheaf.

Collard, R. (2014). Iraq's Battleground Dams are key to saving the country from ISIS. *Time*, 8 September.

Cook, C., & Bakker, K. (2012). Water security: Debating an emerging paradigm. *Global Environmental Change, 22*(1), 94–102.

Cotula, L. (2011). *Land deals in Africa: What is in the contracts?* London: IIED.

Cotula, L. (2012). The international political economy of the global land rush: A critical appraisal of trends, scale, geography and drivers. *Journal of Peasant Studies, 39*(3–4), 649–680.

Darnton, J. (1993). Croats are easing the danger of damaged dam's collapse. *New York Times*, January 31, from http://www.nytimes.com/1993/01/31/world/croats-are-easing-the-danger-of-damaged-dam-s-collapse.html. Accessed 10 Mar 2015.

Davis, J. (2004). Corruption in public service delivery: Experience from South Asia's water and sanitation sector. *World Development, 32*(1), 53–71.

Deininger, K. W., & Byerlee, D. (2011). *Rising global interest in farmland: Can it yield sustainable and equitable benefits?* Washington, DC: World Bank Publications.

European Commission. (2015). *Environmental crime*. From http://ec.europa.eu/environment/legal/crime/. Accessed 20 Feb 2015.

Falkenmark, M. (1990). Global water issues confronting humanity. *Journal of Peace Research, 27*(2), 177–190.

Falkenmark, M., Lundqvist, J., & Widstrand, C. (1990). Coping with water scarcity: Implications of biomass strategy for communities and policies. *International Journal of Water Resources Development, 6*(1), 29–43.

FAO AQUASTAT. (2010). *AQUASTAT database*. From http://www.fao.org/nr/water/aquastat/dbase/AquastatWorldDataEng_20101129.pdf. Accessed 26 Feb 26 2015.

Farmer, L. (2008). Crime, definitions of. In *The New Oxford Companion to Law*. Oxford University Press, from http://www.oxfordreference.com/view/10.1093/acref/9780199290543.001.0001/acref-9780199290543-e-527. Accessed 24 Feb 2015.

Foran, C. (2014). Drought is taking California back to the wild, wild West. *National Journal*, November 10, from http://www.nationaljournal.com/energy/drought-is-taking-california-back-to-the-wild-wild-west-20141110. Accessed 5 Mar 2015.

Franz, A. (2011). Crimes against water: Non-enforcement of state water pollution laws. *Crime, Law and Social Change, 56*(1), 27–51.

Gleick, P. H. (2006). Water and terrorism. *Water Policy, 8*(6), 481–503.

Gleick, P. H. (2013). *Water conflict chronology*. Oakland: Pacific Institute. From http://www2.worldwater.org/conflict/list/. Accessed 24 Feb 2015.

Goldenberg, S. (2014, February 9). Why global water shortages pose threat of terror and war. *The guardian*. Accessed 10 Mar 2015. From: http://www.theguardian.com/environment/2014/feb/09/global-water-shortages-threat-terror-war

Gorantiwar, S. D., & Smout, I. K. (2005). Performance assessment of irrigation water management of heterogeneous irrigation schemes: 1. A framework for evaluation. *Irrigation and Drainage Systems, 19*(1), 1–36.

Grey, D., & Sadoff, C. W. (2007). Sink or swim? Water security for growth and development. *Water Policy, 9*(6), 545.

GWP. (2010). *What is water security?* URL: http://www.gwp.org/The-Challenge/What-is-water-security/. Accessed 13 July 2016.

Haas, L. J., Mazzei, L., & O'Leary, D. (2010). *Lesotho highlands water project*. Washington, DC: Lesotho Highlands Water Project, World Bank.

Hauge, W., & Ellingsen, T. (1998). Beyond environmental scarcity: Causal pathways to conflict. *Journal of Peace Research, 35*, 299–317.

Heidenheimer, A. J. (1989). Perspectives on the perception of corruption. In *Political corruption: A handbook* (pp. 149–163). New Brunswick: Transaction Publishers.

Holzman, D. C. (2011). Methane found in well water near fracking sites. *Environmental Health Perspectives, 119*(7), a289.

Howarth, R. W., Ingraffea, A., & Engelder, T. (2011). Natural gas: Should fracking stop? *Nature, 477*(7364), 271–275.

ICRC. (2015). *Bled dry: How war in the Middle East is bringing the region's water supplies to breaking point, ICRC Report*. Geneva: International Committee for Red Cross.
Jain, A. K. (2001). Corruption: A review. *Journal of Economic Surveys, 15*(1), 71–121.
Jägerskog, A., Swain, A., & Öjendal, J. (2014). *Water security: Vol. 1: Origin and foundations*. London: Sage.
Johnston, M. (1997). Public officials, private interests, and sustainable democracy: When politics and corruption meet. *Corruption and the Global Economy, 83*, 67.
Kelley, C. P., Mohtadi, S., Cane, M. A., Seager, R., Kushnir, Y. (2015). Climate change in the Fertile Crescent and implications of the recent Syrian drought. In *Proceedings of the National Academy of Sciences (PNAS)* of the United States of America.
King, L. W. trans. (1910). *Code of Hammurabi, the ancient history source book*. http://legacy.fordham.edu/halsall/ancient/hamcode.asp. Accessed 13 July 2016.
Klare, M. (2001). *Resource wars: the new landscape of global conflict*. New York: Macmillan.
Land Matrix. (2015). *The online public database on land deals*. 2015, from http://landmatrix.org/en/. Accessed 20 Feb 2015.
Lundqvist, J. (2000). A global perspective on water and the environment. *Physics and Chemistry of the Earth, Part B: Hydrology, Oceans and Atmosphere, 25*(3), 259–264.
MacInnes, M. (2012). *Corruption and large-scale land acquisitions: An analysis of the role high-level corruption plays in enabling elite capture of land*. Paper presented at the international conference on Global Land Grabbing II, October 17–19, 2012. Cornell University, Ithaca, NY, from http://www.cornell-landproject.org-/download/landgrab2012papers/macinnes.pdf. Accessed 26 Feb 2015.
Metha, L., Veldwisch, G. J., & Franco, J. (Ed.). (2012). *Special issue: Water grabbing? Focus on the (re)appropriation of finite water resources*. From http://www.water-alternatives.org/index.php/archive-toc/1881-articles-toc/vol5/213-issue5-2#sthash.xEWbh7GE.dpuf. Accessed 26 Feb 2015.
Milly, P. C. D., Betancourt, J., Falkenmark, M., Hirsh, R. M., Kundzewicz, Z. W., Lettermaier, D. P., & Stouffer, R. J. (2008). Stationarity is dead: Whither water management? *Science, 319*, 573–574.
Momtaz, S. (2002). Environmental impact assessment in Bangladesh: A critical review. *Environmental Impact Assessment Review, 22*(2), 163–179.
Nhantumbo, I., Salomão, A. (2010). *Biofuels, land access and rural livelihoods in Mozambique*, London: IIED, from http://pubs.iied.org/12563IIED.html. Accessed 26 Feb 2015.
Paukste, A. (2004). The definition of terrorism. *Jurisprudencija, 59*, 68–80.
Paul, S. (1995). *A report card on public services in Indian cities: A view from below*. Bangalore: Public Affairs Centre.
Peters, J. G., & Welch, S. (1978). Political corruption in America: A search for definitions and a theory, or if political corruption is in the mainstream of American politics why is it not in the mainstream of American politics research? *The American Political Science Review, 72*(3), 974–984.
Polack, E., Cotula, L., & Côte, M. (2013). *Accountability in Africa's land rush: What role for legal empowerment?* London/Ottawa: IIED/IDRC.
Prüss-Ustün, A., Bartram, J., Clasen, T., Colford, J. M., Cumming, O., Curtis, V., & Cairncross, S. (2014). Burden of disease from inadequate water, sanitation and hygiene in low-and middle-income settings: A retrospective analysis of data from 145 countries. *Tropical Medicine & International Health, 19*(8), 894–905.
Punch, M. (2000). Police Corruption and its prevention. *European Journal on Criminal Policy and Research, 8*(3), 65–90.
Ray, I. (2011). Farm-level incentives for irrigation efficiency: Some lessons from an Indian canal. *Journal of Contemporary Water Research and Education, 121*(1), 10.
Roebuck, J. B., & Barker, T. (1974). A typology of police corruption. *Social Problems, 21*(3), 423–437.
Ruby, C. L. (2002). The definition of terrorism. *Analyses of Social Issues and Public Policy, 2*(1), 9–14.

Rulli, M. C., Saviori, A., & D'Odorico, P. (2013). Global land and water grabbing. *Proceedings of the National Academy of Sciences, 110*(3), 892–897.

Smaller, C., & Mann, H. L. (2009). *A thirst for distant lands: Foreign investment in agricultural land and water*. Winnipeg: International Institute for Sustainable Development-Institut international du dévelopement durable.

Swain, A. (2004). *Managing water conflict: Asia, Africa and the Middle East*. London: Routledge.

Swain, A. (2009). The Indus II and Siachen peace park: Pushing India-Pakistan peace process forward. *The Round Table: The Commonwealth Journal of International Affairs (London), 98*(404), 569–582.

Swain, A. (2012). *Understanding emerging security challenges: Threats and opportunities*. London: Routledge.

The Guardian. (2015). Fracking ban forcing some New York towns to consider redrawing state lines. *The Guardian*, February 28, from http://www.theguardian.com/us-news/2015/feb/28/new-york-fracking-ban-upstate-talk-seceding-pennsylvania. Accessed 3 Mar 2015.

Transparency International. (2008). *Global corruption report 2008- Corruption in the water sector*. Cambridge: Transparency International/Cambridge University Press/Water Integrity Network.

UNCTAD. (2009). *World investment report*. Geneva: United Nations Press.

UNESCO. (2012). *Freedom of information in Africa*. Paris. From www.unesco.org/new/en/communication-and-information/freedom-of-expression/freedom-of-information/foi-in-africa. Accessed 3 Oct 2015.

US State of California. (2015). *California drought*. From http://ca.gov/drought/. Accessed 3 Mar 2015.

VanderMeer, C. (1971). Water thievery in a rice irrigation system in Taiwan. *Annals of the Association of American Geographers, 61*(1), 156–179.

Vargas-Hernández, J. G. (2009). *The multiple faces of corruption: Typology, forms and levels*. Retrieved Febuary 25, 2015, from http://ssrn.com/abstract=1413976 or http://dx.doi.org/10.2139/ssrn.1413976.

Weber, M. (1964). *Economy and society: An outline of interpretive sociology*. Berkeley: University of California Press.

Williams, G. (1955). The definition of crime. *Current Legal Problems, 8*, 107.

Witter, S. G., & Whiteford, S. (1999). Water security: the issues and policy challenges. *International Review of Comparative Public Policy, 11*, 1–25.

Chapter 6
Water Security Is Job Security: Water as an Enabler for Livelihoods

Sarah Dickin and Luca Di Mario

Abstract Water flows through all aspects of our lives, sustaining ecosystems, meeting household needs, and critically, creating opportunities for a range of productive activities. As pressures on the environment increase, understanding the role of water in supporting livelihoods is essential for ensuring the sustained wellbeing of human populations and ecosystems. This chapter examines a variety of pathways through which water of varying qualities and quantities enables livelihoods and incomes. Examples describing the role of water in securing jobs across a range of sectors are explored, including manufacturing industries, energy and agriculture. In particular, this chapter provides insight on the role of wastewater, an increasingly important resource, in providing jobs in growing cities worldwide. Barriers to securing water resources that impede economic opportunities and development, as well as emerging opportunities to reduce these obstacles are presented.

Keywords Water security • Water povery • Livelihoods • Water jobs • Wastewater • Wastewater agriculture

6.1 Introduction

Access to water resources enables a society to develop and prosper. A strong impetus for achieving water security is its importance to a wide range of livelihoods and futures. Many sectors are dependent on reliable access to water, including agriculture and food production, manufacturing, energy, healthcare, transport and tourism. Agriculture is the largest user of freshwater, with irrigation comprising 70% of withdrawals globally, while industry constitutes approximately 20% of the world's freshwater use (WWAP 2012). In addition to the large quantities of water required by these sectors, water security necessitates an adequate level of water quality to

S. Dickin (✉)
Stockholm Environment Institute, Stockholm, Sweden
e-mail: sarah.dickin@sei-international.org

L. Di Mario
Asian Development Bank, Manila, Philippines

D. Devlaeminck et al. (eds.), *The Human Face of Water Security*, Water Security in a New World, DOI 10.1007/978-3-319-50161-1_6

enable livelihoods, ensure wellbeing of human populations and maintain the integrity of aquatic ecosystems. In short, water security is the provision of the right amount of water at the right quality.

Regional differences in economic and environmental conditions play a role in determining livelihood strategies. For instance, farmers growing water-intensive crops such as soybean are often situated in rain-fed areas and hydropower production is concentrated in areas with significant surface water resources. In many regions livelihood strategies are strongly influenced by water stressed conditions and variable water supplies. In low-income countries where agriculture contributes to approximately 15% of overall GDP, compared with 1% in high-income countries, livelihoods are particularly vulnerable to water scarcity and variability (Amarasinghe and Smakhtin 2014). For instance, in Sub-Saharan Africa over 70% of the population depends on subsistence rain-fed agriculture for their livelihoods and has limited access to economic or technical 'safety nets' (IWMI 2014). This underlines the precarious dependence of many rural based economies on water for livelihoods. In high-income countries, water withdrawals within the industrial sectors are over 40%, compared to 5% in some low income countries (WWAP 2012). However, shifts towards greater industrial production do not change our dependence on water resources to provide jobs and livelihoods. Many industries such as manufacturing and mining are critically dependant on access to water resources; moreover, they often require higher quality water than agriculture (e.g. food production and thermal cooling for energy production).

While we increasingly depend on water resources to ensure livelihoods, many productive activities overlook environmental needs for water. For instance, development of water resources in Kimani catchment, Tanzania, has resulted in reliable water supplies and improvements in livelihoods and land-holding size, but has reduced environmental flows to the Usangu wetland downstream (Franks et al. 2013). However, provision of adequate water to nature by incorporating environmental flows into management activities can have important positive impacts, such as diversifying livelihoods strategies and increasing capacity to cope with variability (Schlüter et al. 2013).

6.1.1 Water Poverty and Water Wealth

Maximizing the productive potential of water, while balancing societal and environmental considerations, is a significant challenge for reducing poverty and securing sustainable livelihoods. 'Water poverty' is a term used to describe the limiting of livelihood options and assets due to a lack of sustainable water resources (Kemp-Bendict et al. 2009). It is linked to both the properties of the water resource itself and the capacity of a community to efficiently and effectively use it (Sullivan et al. 2003). However, it is important to recognize that the causes of poverty and their relationship with water resources are very complex, including important environmental and socio-economic components (Namara et al. 2010).

Environmental dimensions of water-related poverty are influenced by physical water availability. This depends on hydrological conditions, including absolute amounts available for use, the distribution of those resources, and their variability, including floods and droughts that contribute to fluctuations. Socio-economic conditions influencing water-related poverty include the availability of infrastructure, institutions and capacities to use water effectively and efficiently (Grey and Sadoff 2007). For instance, in Sub-Saharan Africa, high economic water scarcity is prevalent due to a lack of resources to obtain water for productive use (WWAP 2012). Water resources must be transformed into opportunities in order to enable livelihoods, and systems that can create more benefits with a certain amount of water, such as through the use of technologies, have higher water productivity. These pathways may be limited by a range of barriers, including a lack of access to assets such as financial or human capital, and policy and institutional constraints, such as weak governance or social inequities (Nicol 2000). In the face of these challenges, securing adequate water quality and quantity enables a range of livelihoods options and thus is critical for poverty alleviation and achieving 'water wealth'.

6.1.2 Threats to Water Security as an Enabler for Livelihoods

The growth of human populations together with processes of economic globalisation have significantly changed patterns of production, consumption, and use of environmental services. Human pressures on the environment are transforming earth's life supporting systems, including the climate, nutrient cycles, biodiversity and ecosystem functions, which some have termed 'overstepping planetary boundaries' (Rockström et al. 2009).

These global changes greatly impact the hydrological cycle and in response are impacting the availability for water resources needed to sustain livelihoods. Watersheds are receiving considerable input of contaminants associated with development activities, including excessive nutrients from agricultural runoff and industrial wastewater, which have consequences for water quality and the integrity of aquatic ecosystems. For instance, excess nutrients create hypoxic conditions that alter ecosystem functioning and may damage water-based livelihoods such as fisheries (Díaz and Rosenberg 2011). Land use intensification, including deforestation and extensive agricultural production impact soil quality, and can affect the water balance of a watershed (Schilling et al. 2008). This degradation of ecosystems and their services especially jeopardizes the livelihoods of poor people, who rely on their provisioning services including food and fuel, especially during income shocks (Fisher et al. 2013). In addition, growing abstractions of water and high rates of groundwater use are occurring, in some cases where aquifer resources are finite. This overexploitation threatens many forms of livelihoods relying on sustainable water resources, for instance even the current use of quota-based water policies are not expected to protect the aquifer in the Upper Guadiana basin, Spain, and will result in income losses for rural communities (Várela-Ortega et al. 2011).

Additionally, climate change has direct consequences for water security, further exacerbating water stressed areas and bringing more water to flood prone areas (UN Water 2010). More intense extreme weather events present increased risks to infrastructure, for instance flooding is causing acidification of estuarine waters and increased disease risks for the A$120 oyster aquaculture industry in Australia (Davidson et al. 2013). Flooding and poor management responses in Thailand in 2011 led to billions in losses due to impacts on global supply chains (Grey et al. 2013). Critically, the increasingly unpredictable nature of precipitation patterns will most impact people without the ability to store water to buffer for these changes, such as farmers planting rainfed crops (Namara et al. 2010; Hall et al. 2014). As climate change increasingly contributes to a loss of stationarity in the hydrological cycle, it may become increasingly difficult to rely on many existing forms of job security, while other opportunities will emerge in sectors such as disaster risk management.

While providing significant gains to some populations, global environmental change is impeding the livelihoods of groups who contributed little to the dominant production patterns responsible for these outcomes. These disproportionate impacts show the need to promote water use in ways that allow sustainable livelihoods strategies and equitable growth.

6.2 Diverse Livelihoods Rely on Water Security

Water is essential for most forms of production and creates economic opportunities across a range of sectors (Grey and Sadoff 2007). In this section a range of examples are explored that focus on different pathways through which water security enables livelihoods in various economic and geographic contexts. These examples touch on the role of water in various industries, including energy generation and manufacturing, water-based livelihoods such as fisheries, and agricultural production. Across these cases, the inextricable links between secure water resources and the creation of sustainable jobs is emphasized.

6.2.1 Water as an Enabler for Industry

Global freshwater abstraction for industrial use totals approximately 731 km^3/year according to the AQUASTAT database, and is closely linked to a region's economic development (FAO-AQUASTAT 2012). Manufacturing is an important user of freshwater resources and consumes 27% of groundwater withdrawals (Döll et al. 2012). The Great Lakes Basin, which accounts for around 20% of the world's surface freshwater resources, supports a range of industries, including manufacturing, large shipping routes, and hydropower facilities, and contributes 5 trillion USD to the economy and 39.3 million USD in private employment in the United States and

Canada (WBC 2013). In terms of GDP, this would represent the world's 4th largest economy, greater than that of Germany, and highlights the immense importance of Great Lakes water security to economic activities. Manufacturing contributes more than five million jobs in the region, comprising many water intensive types of production including paper, food and beverages, metals, chemicals including pharmaceuticals, and wood products (WBC 2013). Closely connected to manufacturing industries and agriculture, the cargo and shipping industry which transports bulk goods in the Great Lakes and St. Lawrence River is dependent on water levels in the basin (Krantzberg and De Boer 2008). In this productive region, human development of these water resources, including construction of dams, river diversions, and channel modifications has ensured the water security required by many of these sectors. At the same time many of these industrial manufacturing activities threaten water security in the Great Lakes. The positive impact on labour markets and national economies and must be balanced with steps to ensure current and future ecosystem and human water needs.

In some regions growing demands for water are increasing conflict between alternative uses. In Bangladesh, groundwater use has led the rapid growth of the textile industry and significant job creation. In 2010 garment exports comprised a critical 82% of total merchandise exports from Bangladesh (Ahmed et al. 2014). In addition to contributing approximately 3.6 million jobs, this industry has increased employment opportunities for women in the formal workforce, employing 85% women (WWF 2014). Washing, dyeing and finishing consume 1.5 trillion litres of water, and produce significant volumes of highly polluted wastewater, particularly due to textile dyeing effluent, reducing the quality of water for other uses. In addition, overdrawing groundwater aquifers threatens these livelihoods. In Dhaka overuse has led to a 20 m drop in the water table in the last decade (Chowdhury 2010).

Fisheries are a globally important industry and support the livelihoods of an estimated 520 million people, while also providing at least 20% of the protein intake of 1.5 billion people (FAO 2009). In some regions, such as India, Cambodia and Ghana, fisheries and post-harvest related activities are a key livelihood strategy for women, where they comprise approximately half of the sector's workforce (Badjeck et al. 2010). However, fisheries are strongly linked to water quality and ecosystem integrity, which enable the use of this economically important natural resource. At a household and local level, small-holder fisheries are intensely dependent on healthy aquatic ecosystems to provide livelihoods, emphasizing their importance to water security. In Malawi fisheries provide employment to around 60,000 people directly, and around 450,000 people indirectly, but are negatively impacted by competing uses of lake catchments (Jamu et al. 2011). In Chilika lagoon, in the Bay of Bengal, India more than 400,000 caste-based fishers have created and sustained livelihoods for decades, but are being threatened due to development of shrimp aquaculture (Nayak et al. 2014). Large-scale fisheries have also been impacted by over-use, such as the collapse of cod fisheries in eastern Canada in the early 1990s. In 1993 a moratorium on cod fishing directly impacted 30,000 livelihoods, with indirect impacts on livelihoods approximately 10 times that number (Milich 1999).

While the tourism industry is an important source of employment opportunities, it is increasingly recognized as a large water user. Significant water inputs are needed for tourist accommodations and the industry often depends on the integrity of aquatic ecosystems for recreational activities. Due to the high value of the tourism industry these water uses may outcompete other economic sectors for water supplies, but are likely unsustainable in some regions. For instance, Zanzibar, Tanzania, is a water-scarce island relying on freshwater lenses that has experienced a large growth in tourism and thus an increase in water use that is lowering the groundwater table, deteriorating groundwater quality, and causing saltwater intrusion (Gössling 2001). Continued unsustainable water withdrawals in Zanzibar are expected to impact the tourist industry as well as the livelihoods of local populations. Tourism in the Great Barrier Reef in Australia is particularly vulnerable to changes in water quality, which may increase disease risk and reduce biodiversity with consequences for the tourism industry, which contributes around 5.7 billion AUD to the Australian economy and approximately 70,000 jobs (Marshall et al. 2010; Deloitte Access Economics 2013). Degraded water quality and ecosystem functioning can result in expensive restoration efforts, such as in the Everglades National Park where a 13.5 billion USD 30–40 year plan is underway to restore the hydrologic properties of the ecosystem (NRC 2014).

6.2.2 (Waste)water as an Enabler for Agriculture

Pressures on water resources in the agriculture sector are increasing in line with growing populations and changing preferences in diets, and a rising demand for biofuels. In many low-income countries water security is critically connected to agricultural production, as this represents the main source of livelihoods. Rainfed agriculture comprises 80% of cultivated land worldwide, producing 60% of crops (Wani et al. 2009). Even in regions where withdrawals for agriculture are generally small, reliance on rain presents risks for drier catchments and seasons, which may be exacerbated by climate change. Many countries have invested in infrastructure to ensure agricultural water security, with 30–40% of irrigation water stored behind dams, which secures agricultural livelihoods against variable rainfall (World Commission on Dams 2000).

Today, the proportion of people living in cities accounts for more than 50% of the world's population, up from 34% in 1960, resulting in higher demand for food and water resources (UN 2014). Migration towards cities to obtain better livelihood opportunities including jobs, services (e.g. health, education) and protection from natural disasters, such as droughts, continue to drive urbanization globally (Seto 2011). With these rapid changes to city landscapes, urban and peri-urban agriculture (UPA) has become an important income generating activity and a way of adapting to variable water resources. In addition, the current contribution of UPA to global food security is non-trivial. It is estimated that 68 million hectares are cultivated under UPA including both irrigated and rainfed areas, which is larger than of all rice growing areas in South Asia, and the trend continues to move upwards (Thebo et al. 2014).

With the expansion of urban settlements, wastewater is a growing and critically important source of water security for livelihoods. Concentrated development with concurrent increases in pollution loads that are released into the natural environment, as well as a higher demand for resources places substantial pressure on water supplies and waste treatment infrastructure in cities (and in the surrounding areas) (Corcoran et al. 2010). This especially impacts the availability of freshwater and as a response to this scarcity and the rise of marginal quality water flows, urban and peri-urban farmers often shift to using wastewater. About 20 million hectares (nearly 10% of total irrigated land) are estimated to be irrigated with wastewater (Jiménez et al. 2010). In one survey, four out of every five cities in developing regions were using wastewater for irrigation, predominantly for local market sale and livelihood support (Raschid-Sally and Jayakody 2009). The use of wastewater is driven by its (van der Hoek et al. 2002; Hussain et al. 2002):

- Greater physical availability and reliability – as the only secure water source in some areas;
- Low cost – free or cheaper than other water sources; and
- Additional benefits due to nutrients and organic matter content.

While the health risks must be taken in to consideration (Dickin et al. 2016), the use of wastewater has several important implications for enabling livelihoods. The most important value added benefit associated with wastewater irrigation is attributable to its constant and reliable supply, in other words water security (Fig. 6.1). Farmers irrigating with wastewater are able to crop during the dry-season and achieve year-round production, which increases revenues per hectare and gross annual incomes (Danso et al. 2014). For example, evidence from Ghana shows that additional revenues from dry-season cultivation only (with wastewater) contribute an extra 100% of income compared with rainfed agriculture systems in the same area (Danso et al. 2014). These benefits are evident from revenues generated by different farming systems in Ghana shown in Table 6.1.

A similar story can be inferred from the evidence collected by Buechler et al. (2002) along the Musi River in India, where about 50,000 people were estimated to

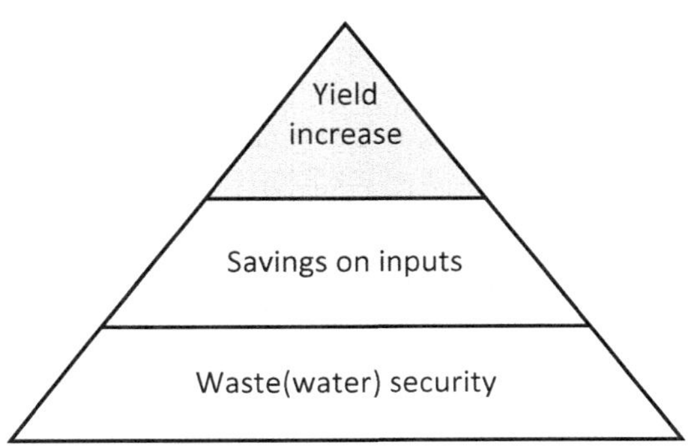

Fig. 6.1 The benefits of wastewater for urban and peri-urban farmers

Table 6.1 Revenue generated in different farming systems in 2002

Location	Farming system	Typical farm size (ha)	Net revenue (USD) per actual farm size per year (range)
Rural/ peri-urban	Rainfed maize or maize/cassava	0.5–0.9	200–450[a]
Urban	Year-round irrigated vegetable farming (lettuce, cabbage, spring onion)	0.05–0.2	400–800

Danso et al. (2002)

[a]These are typical values; subsistence production has been converted to market values. If farmers use parts of their maize and cassava harvest for home consumption, the actual net income would be lower

be directly or indirectly relying on wastewater. The Musi River relies on the wastewater flows from Hyderabad, particularly during the dry season when the available freshwater dries up. Farmers irrigating leafy vegetables and fodder grass with wastewater generated income of around Rs. 40,000 and Rs. 135,000 per hectare per year respectively.

This increase is even greater in extremely dry contexts such as in Lima, Peru, where rainfall is almost absent (6–13 mm per year) and urban and peri-urban farmers are able to cultivate vegetables and other crops nearly exclusively relying on wastewater availability (Merzthal and Bustamante 2008). Wastewater reuse in agriculture, if undiluted, may bring further benefits due to nutrients contained (e.g. nitrogen and phosphorous), which when recognised by farmers, may substantially offset the amount of fertilisers and input cost. For example, in Pakistan, evidence indicated a higher gross margin (Rs. 8369/ha) for farmers irrigating with wastewater vs. others irrigating with freshwater from canals (Rs. 1786/ha). This was mainly due to lower inputs costs (e.g. fertilizers) and higher agricultural yield (Hussain et al. 2002).

Therefore, the contribution of wastewater to livelihoods is multi-faced (Fig. 6.2). This includes direct benefits such as greater yields, outputs and revenues at lower costs, and hence higher incomes, better returns for labour and greater access to agricultural assets. There are also notable indirect benefits such as greater business productivity, industrialization and market participation. These benefits are not restricted to agricultural activities in the peri-urban interface, rather they stretch beyond the agricultural sector. However, reliable economic comparisons are rare as farmers in rural areas are commonly used for comparison for studies on health and livelihood improvements (Dreschel et al. 2015). Further work is needed to provide additional evidence on the indirect benefits of wastewater use, such as ecosystem protection, health related risks and benefits, and climate change adaptation.

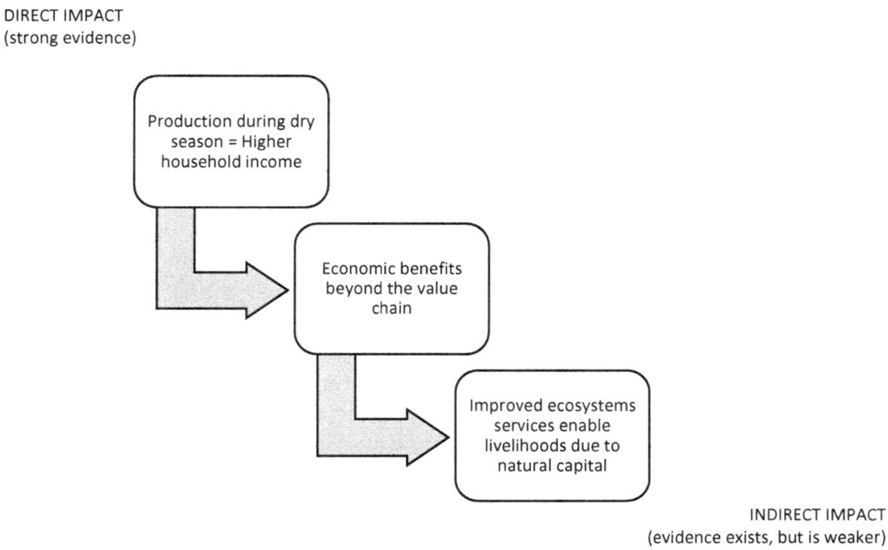

Fig. 6.2 Direct and indirect benefits of wastewater

6.3 Water Insecurity as a Barrier to Livelihoods

A range of social, economic, political and physical factors that create conditions of water insecurity can contribute to a loss of livelihoods, including environmental degradation, infrastructure needs, limited economic capital and ineffective governance mechanisms which impact access to and management of resources. Identifying the types of constraints that limit livelihoods sheds light on appropriate solutions to improve water security and highlights pathways to ensure an equitable balance is achieved between different users. Challenges associated with water governance and pollution of water resources are discussed as examples of obstacles to securing livelihoods.

Good water governance plays a key role in balancing trade-offs to ensure water security for both human and environmental needs (Bakker 2012). While water productivity decreases poverty, the cost of water or rights to access water for income generating activities may represent barriers to livelihoods and disproportionately impact the poor (Namara et al. 2010). Equitable management of water resources is essential for ensuring that some livelihoods are not secured at the cost of others, however without adequate steps to manage competing demands this is a prevalent challenge due to formal and informal power imbalances. For instance, in West Java, Indonesia, textile factories seized water rights of smallholder farmers, while in the Philippines farmers using water in irrigation schemes experienced restrictions due to municipal uses (Watkins et al. 2006). In particular, transboundary water agreements may result in uneven impacts on some livelihoods. Projects designed to secure water resources in one region can have widespread impacts on downstream populations, including increased water scarcity. For instance, hydropower generation can provide large contributions to water security and livelihoods, but may negatively impact some groups that depend on the river systems for their livelihoods, as this development is often linked to river fragmentation and ecosystem disruption (WCD 2000). In Central Asia, large upstream water resources allow Tajikistan to generate electricity, but this has had implications for downstream livelihoods in countries practicing water-intensive cotton, rice and wheat irrigation. In the lower Mekong delta, more than 45 million people rely on the river system for their livelihoods, which will be impacted by ongoing dam construction that will alter river ecology and income-generation from fisheries in particular (WWAP 2014). A gender gap between men and women's access to water for productive purposes is another challenge for enabling livelihoods, impacting the economic opportunities for many women and their households. Closing the agricultural gap by increasing access to assets, inputs and services including water resources would increase yields of women's farms by 20–30% (FAO 2011). A lack of access to domestic water and sanitation supplies also has direct consequences for women in many countries, as this requires a substantial time input, reducing time available for productive activities that generate livelihoods.

Pollution of water resources due to contaminant input and agricultural run-off into surface and groundwater supplies severely threatens human health and income generating activities of many groups (Millennium Ecosystem Assessment 2005). For instance, contamination of the Ogoniland wetlands in the Niger Delta due to oil spills associated with the fossil fuel industry impacted the high biodiversity, provisioning and regulating services of the ecosystem. This resulted in severe implications for indigenous groups depending on these services for their livelihoods (UNEP 2011). In the case of urban and peri-urban agriculture, wastewater contains nutrients that are useful for supporting productive livelihood activities, as well as potentially harmful chemicals and pathogens. In many cases highly polluted wastewater may be the only available water source to farmers, however the benefits from using this water may be overshadowed by the possible health risks for farmers and consumers (Mara and Horan 2003; WHO 2006; Drechsel et al. 2009). Critically, the impacts of many long-lasting pollutants, such as persistent organic pollutants or heavy metals, threaten existing livelihoods as well as those of future generations.

6.4 Implications for Water and Livelihoods in a Changing World

Ensuring water security to enable livelihoods will require mitigating risks, removing barriers and identifying appropriate interventions to maintain water use at sustainable levels. Specific strategies will differ between geographic regions with varying hydrological conditions; between low, middle and high incomes countries where dominant water uses for supporting livelihoods vary between industries and agricultural applications; and between particular circumstances such as water insecurity arising from natural hazards or conflicts. For instance, following conflicts or disasters water security has been recognized as a first step in restoring livelihoods, economic development and human wellbeing (UN Water 2013). In some cases, systemic changes to the environment will irrevocably impact jobs and livelihoods associated with water today, and require the identification of new opportunities. In addition to water needs, it is important to note that many sectors are also facing greater competition for land and energy resources, and solutions that address the nexus of these of issues are needed. Steps to address threats to water security must be taken proactively to avoid expensive consequences and remediation down the road (Vörösmarty et al. 2010).

Innovations in policy options are needed to ensure sustained water security for livelihoods in changing environments. For instance, the OECD water governance initiative has recognized the important role of water in job creation and economic development. A key objective of this initiative was to provide policy guidelines to enhance management of water resources by engaging decision-makers at all levels (OECD 2015). Research into gender barriers in agriculture livelihoods has identified options such as collective ownership of land or irrigation water resources to reduce the gender gap. This allows women farmers to overcome challenges linked to lack of access to agricultural assets and scale constraints (Sugden et al. 2014). Increasing access to water through 'smart-subsidies' that allow investment in rainwater harvesting may be appropriate in some contexts. Given the great importance of wastewater (treated and untreated) for many livelihoods, policy solutions should look towards integrating safer reuse alternatives and protocols, rather than adopting strict guidelines or bans. Full-treatment approaches for wastewater may not be feasible to implement or enforce and may be detrimental for livelihoods, at least in the short run, and produce counterproductive effects on human health (e.g. undernutrition) as often the supply of 'salad greens' for whole cities depends on wastewater use (Drechsel et al. 2006; Drechsel et al. 2009). Sanitation safety planning is increasingly promoted by the WHO to address safety and risk issues from farm to fork, protecting both consumers and livelihoods.

Technology advances, such as the use of alternative raw water sources and improvements in efficiencies will be needed in water stressed regions (Falkenmark 2013). These investments also have the substantial potential for creation of jobs in sustainable water management. For instance, upgrading conventional stormwater infrastructure to alternative 'green infrastructure' that allows improved infiltration, evaporation and reuse requires substantial investment. Moore et al. 2013 identified 136 occupations associated with attaining more sustainable water usage, including in agricultural, domestic and commercial sectors, restoration and remediation, alternative water sources, and stormwater management. In the United States, 10–15 jobs are created per $1 million invested in alternative water supplies, 14.6 in agricultural efficiency and quality, 12–22 in urban conservation and efficiency, 5–20 in stormwater management; and 10–72 jobs per $1 million invested in restoration and remediation (Moore et al. 2014). In low-income countries, developing water systems that can provide productive and domestic water supplies may provide synergistic effects and encourage additional income generating activities (Van Houweling et al. 2012). For example, challenges with rain-fed agriculture that relies on increasingly uncertain water supplies have led to efforts in livelihood diversification. Households that have access to a form of productive water can diversify activities. For instance, domestic water supplies were used for pig-raising, drink stands, small eateries, tea shops, hairdressing shops, and motorbike washing in rural Vietnam (Noel et al. 2010).

In the case of wastewater, better treatment technologies that improve water quality will decrease health and environmental risks. While so far this is only possible in high income countries where taxes and user fees can pay for the services and additional energy inputs needed, wastewater treatment may open up different opportunities for resource recovery (e.g. systemic nutrient, energy and organic matter recovery) which may relieve pressure on aquatic ecosystems, enhance soil quality and reduce greenhouse gas emissions. At different scales this may result in substantial positive benefits to ecosystems and free up ecosystems services (particularly downstream) which can positively impact a range of livelihoods that rely on natural capital such as fishers.

6.4.1 Holistic Solutions to Enable Livelihoods

Managing for water security has implications for many interconnected sectors, providing opportunities for additional dividends on investments. From a 'nexus' perspective, water insecurity is closely related to food insecurity, and has led to combined opportunities to manage these challenges. For instance, in Ethiopia, extreme water stress has led to recurring food shortages and loss of agricultural livelihoods. A 'Productive Safety Net Program' designed to provide livelihood opportunities during these water scarce periods through public works projects including water conservation activities improved groundwater recharge, and thus agricultural productivity and livelihoods were supported (Grosh et al. 2008). Addressing these complex challenges in a holistic manner is important, and there is a need for research to assess competing policy and technology options to improve the ways water is used to support livelihoods. For instance, in the energy sector algal biofuel technology is being developed but is currently too energy and water intensive and may have unknown impacts on coastal ecosystems (Chisti 2013).

Considering the integrated nature of many solutions highlights indirect benefits of water security for livelihoods, as benefits may extend to different scales and socio-economic contexts along the value chain. For instance, investment in safe drinking water and sanitation makes important contributions to productivity and participation in the labour market. Reducing the associated burden of water-related disease decreases lost work due to illness, while also allowing women to devote more time to productive activities and children to attend school, which leads to greater income-generating opportunities. The indirect benefits provided by (waste) water security, may have upstream (e.g. inputs sales and distribution) and downstream (e.g. markets and distributions of products) effects, providing extra jobs and further income generating opportunities in cities and their peripheries (Buechler et al. 2002). Furthermore, freeing up good quality freshwater for other urban uses (e.g. drinking water or industries) in exchange for treated wastewater use in agriculture could enable other livelihoods (Rao et al. 2015). Technological and management choices should consider options that fit emerging local opportunities, such as those that can take advantage of resource recovery. Creating 'win-win-win' scenarios that decrease pollution, provide resources, and decrease or optimise costs and energy inputs enable better conditions to ensure livelihoods and sustainable development in the long-run. At a global level, the UN adopted a series of Sustainable Development Goals (SDGs) in 2015 as part of a 2030 Development Agenda, including a goal to 'Ensure availability and sustainable management of water and sanitation for all.' This goal includes a range of targets related to achieving water and sanitation access, water resources management, and addressing water quality and risks, and has linkages to other goals and targets including disaster risk reduction, health and ecosystem integrity. The SDGs provide a framework to promote the importance of water in achieving social, economic and environmental objectives, and meeting these targets within countries will support a range of livelihoods and job opportunities (Schuster-Wallace et al. 2015).

6.5 Conclusions

Achieving and sustaining water security is likely to be an on-going societal challenge as the world is shaped by a growing array of changes impacting human populations and the environment. Despite these changes water will continue to be a source of livelihoods, economic wellbeing and prosperity. This chapter provides an overview of the importance of water for economic opportunities, providing another dimension to the human face of water security. Diverse examples illustrate livelihoods enabled by continued access to adequate quantities and qualities of water, and demonstrate how water-related livelihoods strategies vary by geographical context.

New challenges to ensuring water security, including climate change and urbanizing populations, are changing the way we think about water, who benefits from it, and its fate after it has been used (Falkenmark 2013). In addition to considering traditional surface and groundwater resources, alternative and emerging ways to secure water are discussed. The use of wastewater for urban and peri-urban farming in growing cities provides insight into new directions that will be required in many regions to enable water-related livelihood strategies. In many cases considering water and livelihoods alone is not enough, and an integrated or 'nexus' approach will be needed to avoid negative impacts on other sectors and ensure goals for sustainable development are met. This chapter describes some of the obstacles to achieving water security that hamper access to livelihoods, as well as examples of policy and technology solutions that have been applied in particular contexts to enable livelihoods. More of these solutions will be needed to provide diverse livelihoods for growing populations, many of whom settle close to water resources and rely on their continued availability for economic opportunities.

Acknowledgements We would like to thank Pay Drechsel and Munir Hanrja for providing insightful comments and suggestions.

References

Ahmed, F. Z., Greenleaf, A., & Sacks, A. (2014). The paradox of export growth in areas of weak governance: The case of the ready-made garment sector in Bangladesh. *World Development, 56*, 258–271.

Amarasinghe, U. A., & Smakhtin, V. (2014). *Global water demand projections: Past, present and future (Vol. 156)*. Colombo: International Water Management Institute (IWMI).

Badjeck, M. C., Allison, E. H., Halls, A. S., & Dulvy, N. K. (2010). Impacts of climate variability and change on fishery-based livelihoods. *Marine Policy, 34*(3), 375–383.

Bakker, K. (2012). Water security: Research challenges and opportunities. *Science, 337*(6097), 914–915.

Buechler, S., Devi, G., & Raschid, L. (2002). Livelihoods and wastewater irrigated agriculture along the Musi River in Hyderabad City, Andhra Pradesh, India. *Urban Agriculture Magazine, 8*, 14–17.

Chisti, Y. (2013). Constraints to commercialization of algal fuels. *Journal of Biotechnology, 167*(3), 201–214.

Chowdhury, N. (2010). Water management in Bangladesh: An analytical review. *Water Policy, 12*(1), 32–51.

Corcoran, E., Nellemann, C., Baker, E., Bos, R., Osborn, D., & Savelli, H. (2010). *Sick water? The central role of wastewater management in sustainable development. A rapid response assessment*. United Nations Environment Programme. United Nations Environment Programme, UN-HABITAT, GRID-Arendal.

Danso, G., Drechsel, P., Wiafe-Antwi, T., & Gyiele, L. (2002). Income of farming systems around Kumasi, Ghana. *Urban Agriculture Magazine, 7*, 5–6.

Danso, G., Hope, L., & Drechsel, P. (2014). Financial and economic aspects of urban vegetable farming. In D. P. & K. B. (Eds.), *Irrigated urban vegetable production in Ghana: Characteristics, benefits and risk mitigation* (pp. 38–50). Colombo: International Water Management Institute (IWMI).

Davidson, J. L., van Putten, I. E., Leith, P., Nursey-Bray, M., Madin, E. M., & Holbrook, N. J. (2013). Toward operationalizing resilience concepts in Australian marine sectors coping with climate change. *Ecology and Society, 18*(3), 4.

Deloitte Access Economics. (2013). *Economic contribution of the Great Barrier Reef, Great Barrier Reef Marine Park*. Townsville: Authority.

Diaz, R. J., & Rosenberg, R. (2011). Introduction to environmental and economic consequences of hypoxia. *Water Resources Development, 27*(01), 71–82.

Dickin, S. K., Schuster-Wallace, C., Qadir, M., & Pizzacalla, K. (2016). A review of health risks and pathways for exposure to wastewater use in Agriculture. *Environmental Health Perspectives*. doi:10.1289/ehp.1509995.

Döll, P., Hoffmann-Dobrev, H., Portmann, F. T., Siebert, S., Eicker, A., Rodell, M., et al. (2012). Impact of water withdrawals from groundwater and surface water on continental water storage variations. *Journal of Geodynamics, 59*, 143–156.

Drechsel, P., Graefe, S., Sonou, M., & Cofie, O. O. (2006). *Informal irrigation in urban West Africa: An overview* (IWMI research report series 102). Colombo: IWMI.

Drechsel, P., Scott, C. A., Raschid-Sally, L., Redwood, M., & Bahri, A. (2009). *Wastewater irrigation and health: Assessing and mitigating risk in low-income countries*. London: Earthscan.
Drechsel, P., Danso, G., & Qadir, M. (2015). Wastewater use in agriculture: Challenges in assessing costs and benefits. In P. Drechsel, M. Qadir, & D. Wichelns (Eds.), *Wastewater: Economic asset in an urbanizing world* (pp. 139–152). Dordrecht: Springer.
Falkenmark, M. (2013). Adapting to climate change: Towards societal water security in dry-climate countries. *International Journal of Water Resources Development, 29*(2), 123–136.
FAO. (2009). *The state of world fisheries and aquaculture*. Rome: FAO.
FAO. (2011). *Women in agriculture: Closing the gender gap for development*. Rome: FAO.
FAO-AQUASTAT. (2012). FAO Global information system on water and agriculture. http://www.fao.org/nr/water/aquastat/maps/index.stm
Fisher, J. A., Patenaude, G., Meir, P., Nightingale, A. J., Rounsevell, M. D., Williams, M., & Woodhouse, I. H. (2013). Strengthening conceptual foundations: Analysing frameworks for ecosystem services and poverty alleviation research. *Global Environmental Change, 23*(5), 1098–1111.
Franks, T., Cleaver, F., Maganga, F., & Hall, K. (2013). Water governance and livelihoods: Outcomes for smallholders on the Usangu plains, Tanzania. *Water Resources and Rural Development, 1*, 1–16.
Gössling, S. (2001). The consequences of tourism for sustainable water use on a tropical island: Zanzibar, Tanzania. *Journal of Environmental Management, 61*(2), 179–191.
Grey, D., & Sadoff, C. W. (2007). Sink or swim? Water security for growth and development. *Water Policy, 9*(6), 545.
Grey, D., Garrick, D., Blackmore, D., Kelman, J., Muller, M., & Sadoff, C. (2013). Water security in one blue planet: Twenty-first century policy challenges for science. *Philosophical Transactions of the Royal Society of London A, 371*(2002), 20120406.
Grosh, M., del Ninno, C., Tesliuc, E., & Ouerghi, A. (2008). *The design and implementation of effective safety nets: For protection and promotion*. Washington, DC: The World Bank.
Hall, J. W., Grey, D., Garrick, D., Fung, F., Brown, C., Dadson, S. J., & Sadoff, C. W. (2014). Coping with the curse of freshwater variability. *Science, 346*(6208), 429–430.
Hussain, I., Raschid, L., Hanjra, M. A., Marikar, F., & Van der Hoek, W. (2002). *Wastewater use in agriculture: Review of impacts and methodological issues in valuing impacts (Working Paper 37)*. Colombo: International Water Management Institute.
International Water Management Institute (IWMI). (2014). IWMI Research East Africa. Accessed online: http://www.iwmi.cgiar.org/Publications/Other/PDF/iwmi_research_in_east_africa-brochure.pdf
Jamu, D., Banda, M., Njaya, F., & Hecky, R. E. (2011). Challenges to sustainable management of the lakes of Malawi. *Journal of Great Lakes Research, 37*, 3–14.
Jiménez, B., Drechsel, P., Koné, D., Bahri, A., Raschid-Sally, L., & Qadir, M. (2010). Wastewater, sludge and excreta use in developing countries: An overview. In P. Drechsel et al. (Eds.), *Wastewater irrigation and health: Assessing and mitigating risk in low-income countries* (pp. 3–27). London: Earthscan-IDRC-IWMI.
Kemp-Benedict, E., Bharwani, S., de la Rosa, E., Krittasudthacheewa, C., & Matin, N. (2009). *Assessing water-related poverty using the sustainable livelihoods framework*. Stockholm: Stockholm Environment Institute.
Krantzberg, G., & de Boer, C. (2008). A valuation of ecological services in the Laurentian Great Lakes Basin with an emphasis on Canada. *American Water Works Association. Journal, 100*(6), 100.
Mara, D., & Horan, N. J. (2003). *Handbook of water and wastewater microbiology*. Amsterdam: Academic Press.
Marshall, N. A., Marshall, P. A., Tamelander, J., Obura, D., Malleret-King, D., & Cinner, J. E. (2010). *A framework for social adaptation to climate change: Sustaining tropical coastal communities and industries*. Gland: International Union for Conservation of Nature.
Merzthal, G., & Bustamante, E. (2008). Using treated domestic wastewater for urban agriculture and green areas: The case of Lima. *Urban Agriculture Magazine, 20*, 7–9.

Milich, L. (1999). Resource mismanagement versus sustainable livelihoods: The collapse of the Newfoundland cod fishery. *Society & Natural Resources, 12*(7), 625–642.
Millennium Ecosystem Assessment. (2005). *Ecosystems and human well-being* (Vol. 5). Washington, DC: Island Press.
Moore, E., Cooley, H., Christian-Smith, J., Donnelly, K., Ongoco, K., & Ford, D. (2013). *Sustainable water jobs: A national assessment of water-related green job opportunities*. Oakland: Pacific Institute.
Moore, E., Cooley, H., Christian-Smith, J., & Donnelly, K. (2014). Sustainable water jobs. In P. H. Gleick (Ed.), *The world's water volume 8: The biennial report on freshwater resources* (Vol. 8). Washington, DC: Island Press.
Namara, R. E., Hanjra, M. A., Castillo, G. E., Ravnborg, H. M., Smith, L., & Van Koppen, B. (2010). Agricultural water management and poverty linkages. *Agricultural Water Management, 97*(4), 520–527.
National Research Council (NRC). (2014). *Progress toward restoring the everglades: The fifth biennial review*. Washington, DC: The National Academies Press.
Nayak, P. K., Oliveira, L. E., & Berkes, F. (2014). Resource degradation, marginalization, and poverty in small-scale fisheries: Threats to social-ecological resilience in India and Brazil. *Ecology and Society, 19*(2), 73.
Nicol, A. (2000). *Adopting a sustainable livelihoods approach to water projects: Implications for policy and practice*. London: Overseas Development Institute.
Noel, S., Phuong, H. T., Soussan, J., & Lovett, J. C. (2010). The impact of domestic water on household enterprises: Evidence from Vietnam. *Water Policy, 12*(2), 237.
OECD. (2015). Stakeholder engagement for inclusive water governance, OECD studies on water. Paris: OECD Publishing. http://dx.doi.org/10.1787/9789264231122-en.
Rao, K., Hanjra, M. A., Drechsel, P., & Danso, G. (2015). Business models and economic approaches supporting water reuse. In D. P., Q. M., & W. D. (Eds.), *Wastewater: Economic asset in an urbanizing world* (pp. 195–216). Dordrecht: Springer.
Raschid-Sally, L., & Jayakody, P. (2009). *Drivers and characteristics of wastewater agriculture in developing countries: Results from a global assessment* (Vol. 127). Colombo: IWMI.
Rockström, J., Steffen, W., Noone, K., Persson, Å., Chapin, F. S., Lambin, E. F., et al. (2009). A safe operating space for humanity. *Nature, 461*(7263), 472–475.
Schilling, K. E., Jha, M. K., Zhang, Y. K., Gassman, P. W., & Wolter, C. F. (2008). Impact of land use and land cover change on the water balance of a large agricultural watershed: Historical effects and future directions. *Water Resources Research, 44*(7).
Schlüter, M., Khasankhanova, G., Talskikh, V., Taryannikova, R., Agaltseva, N., Joldasova, I., et al. (2013). Enhancing resilience to water flow uncertainty by integrating environmental flows into water management in the Amudarya River, Central Asia. *Global and Planetary Change, 110*, 114–129.
Schuster-Wallace, C. J., Sandford, R., Dickin, S. K., Vijay, M., Laycock, K., & Adeel, Z. (2015). *Water in the world we want*. Hamilton: United Nations University Institute for Water, Environment and Health and United Nations Office for Sustainable Development.
Seto, K. C. (2011). Exploring the dynamics of migration to mega-delta cities in Asia and Africa: Contemporary drivers and future scenarios. *Global Environmental Change, 21*, S94–S107.
Sugden, F., Maskey, N., Clement, F., Ramesh, V., Philip, A., & Rai, A. (2014). Agrarian stress and climate change in the Eastern Gangetic Plains: Gendered vulnerability in a stratified social formation. *Global Environmental Change, 29*, 258–269.
Sullivan, C. A., Meigh, J. R., & Giacomello, A. M. (2003). The water poverty index: Development and application at the community scale. *Natural Resources Forum, 27*(3), 189–199.
Thebo, A. L., Drechsel, P., & Lambin, E. F. (2014). Global assessment of urban and peri-urban agriculture: Irrigated and rainfed croplands. *Environmental Research Letters, 9*(11), 114002.
UN Water. (2010). *Climate change adaptation: The pivotal role of water*. New York: United Nations.
UN Water. (2013). *Water security and the global water agenda: A UN-water analytical brief*. Hamilton: UN University.

UNEP. (2011). *UNEP environmental assessment of Ogoniland*. Nairobi: UNEP.

United Nations. (2014). *World urbanisation prospects, the 2014 revision*. New York: United Nations – Department of Economic and Social Affairs, Population Division.

Van der Hoek, W., Hassan, M. U., Ensink, J. H., Feenstra, S., Raschid-Sally, L., Munir, S., et al. (2002). *Urban wastewater: A valuable resource for agriculture: A case study from Haroonabad, Pakistan* (Vol. 63). Colombo: IWMI.

Van Houweling, E., Hall, R. P., Sakho Diop, A., Davis, J., & Seiss, M. (2012). The role of productive water use in women's livelihoods: Evidence from rural Senegal. *Water Alternatives, 5*(3), 658–677.

Varela-Ortega, C., Blanco-Gutiérrez, I., Swartz, C. H., & Downing, T. E. (2011). Balancing groundwater conservation and rural livelihoods under water and climate uncertainties: An integrated hydro-economic modeling framework. *Global Environmental Change, 21*(2), 604–619.

Vörösmarty, C. J., McIntyre, P. B., Gessner, M. O., Dudgeon, D., Prusevich, A., Green, P., et al. (2010). Global threats to human water security and river biodiversity. *Nature, 467*(7315), 555–561.

Wani, S. P., Rockström, J., & Oweis, T. Y. (2009). *Rainfed agriculture: Unlocking the potential* (Vol. 7). Wallingford: CABI.

Watkins, K., Carvajal, L., Coppard, D., Fuentes, R., & Ghosh, A. (2006). *Beyond scarcity: Power, poverty and the global water crisis*. New York: United Nations Development Programme, Palgrave McMillan.

WHO. (2006). *WHO guidelines for the safe use of wastewater, excreta and greywater, wastewater use in agriculture* (Vol. 2). Geneva: WHO.

World Business Chicago (WBC). (2013). *Great Lakes & St. Lawrence region 2013 economy profile update*. Chicago: WBC.

World Commission on Dams (WCD). (2000). *Dams and development: A new framework for decision-making: The report of the world commission on dams*. London: Earthscan.

World Water Assessment Programme (WWAP). (2012). *The united nations world water development report 4: Managing water under uncertainty and risk*. Paris: UNESCO.

World Water Assessment Programme (WWAP). (2014). *The united nations world water development report: Water and energy*. Paris: UNESCO.

WWF. (2014). The imported risk: Germany's water risks in times of globalisation. World wildlife fund. WWF Germany. http://www.wwf.de/fileadmin/fm-wwf/Publikationen-PDF/WWF_Study_Waterrisk_Germany.PDF

Chapter 7
Water Seekers, Carriers and Keepers: The Global and Gender Divide

Alice Bouman-Dentener

Abstract Not having access to sufficient and safe water for basic needs is a feature of extreme poverty. Inadequate water supply and sanitation continues to be the most harmful water risk for people and globally accounts for the largest economic losses. Although gender statistics on water are scarce and scattered, it is safe to say that the larger burden still falls on women and girls, who are the traditional water seekers and carriers in secluded communities and least developed countries, but generally do not have a voice in decision-making concerning water supply and management. It is argued that the water-gender-development nexus (SDG#5 – SDG#6 interface) is a promising and largely untapped connection to reach those furthest behind, in particular through the meaningful involvement of women at all levels and stages of water management processes as called for in Dublin Principle 3 for Integrated Water Resources Management of 1992. Voices of women from Sub-Saharan Africa, a region where water and gender divides are among the highest in the world, illustrate the ingredients and processes of women's empowerment and their inclusion in water governance, and how addressing water in conjunction with gender has a positive and lasting impact on community development as a whole. A comprehensive water-gender-sustainable development strategy gives due consideration to women's civil society, whose potential has been only marginally utilised to date.

Keywords Water • Women • Gender • SDG5 • SDG6 • Poverty • Sustainable development

A. Bouman-Dentener (✉)
DiploriA Sustainable Development Solutions, The Hague, The Netherlands

Women for Water Partnership, The Hague, The Netherlands
e-mail: alice.bouman@gmail.com; alice@diploria.org; hfp@womenforwater.org

D. Devlaeminck et al. (eds.), *The Human Face of Water Security*, Water Security in a New World, DOI 10.1007/978-3-319-50161-1_7

7.1 Introduction

Transforming Our World: the 2030 Agenda for Sustainable Development (General Assembly Resolution 70/1) aims to eradicate poverty in all its forms and dimensions by the year 2030. It is an ambitious agenda that wants to reach the furthest behind first and which emphasises the importance of combatting inequalities, including by promoting gender equality and the empowerment of women and girls to achieve an all-of-society engagement and partnership for sustainable development.

The new Development Agenda includes a dedicated water goal – SDG #6 – underpinning the importance of water security as the foundation and the glue for achieving food and energy security and for building safe and peaceful societies in which all human beings can lead productive lives in harmony with nature. SDG #6 addresses the entire spectrum of water-related risks and includes targets for cooperation, capacity building and community involvement.

Despite the considerable progress made under the Millennium Development Goals (MDGs), today's reality still is that water insecurity keeps millions of people in poverty, continues to obstruct human well-being and seriously threatens the ecosystems that sustain our planet. The World Water Development Report 2015 describes reducing poverty through water management as a useful pro-poor framework for action and points to the links with water quality, access, livelihood opportunities, capacity building and empowerment, water-related disaster prevention and management, and ecosystem management (WWAP 2015).

Water is in everything and when trying to find a way out of poverty, access to sufficient and safe water for basic needs, for both domestic and productive uses, is a basic requirement. According to a recent GWP/OECD report, inadequate water supply and sanitation are the most harmful risks to people and continue to have the greatest economic consequence of all water-related risks, with estimated total economic losses of US$ 260 billion annually; these losses includes the per capita estimates of the value of time spent to fetch water (Sadoff et al. 2015).

There is ample documentation to prove that lack of sufficient and safe water and adequate sanitation facilities is largely a rural and poverty-related phenomenon. In addition to rural/urban and wealth-related disparities, coverage is also much lower for minorities and other disadvantaged groups (The Millennium Development Goals Report 2014). While gender disaggregated data in the water domain are scarce and scattered, it is safe to say that the larger burden still falls on women and girls, who are the traditional water seekers and carriers in the developing world while at the same time being largely excluded from decision-making and water management processes.

7.2 Why Work on the Water-Gender-Development Nexus

Gender statistics on water are among the least available of national-level indicators. The MDG indicators for water have not been disaggregated by sex and 45.2% of countries do not produce any gender statistics related to water (World Water Assessment

Programme 2013). There is general consensus however, that water-related development interventions are not gender-neutral. Priorities for water allocation reflect unequal social status and power relations between women and men, resulting in situations where the water needs for women's domestic and productive roles are often not being adequately covered. There is ample documentation to prove that, when water is not available on the premises – which is the case in the majority of Sub-Saharan African countries - the main burden for collecting water falls on women, while their participation in the management of natural resources such as water is limited (UN DESA 2010). Chapter 6 of the UN Women 2014 global survey on the role of women in development ascertains that the domain of water and sanitation has a particularly strong potential to transform the lives of women and girls (UN Women 2014).

In rural areas of the Least Developed Countries, the gender-water-development interlinkages are particularly strong and connected to agricultural livelihoods. Agriculture is the main consumer of water, accounting for 70% of total water withdrawals globally (FAO 2015). More than 80% of the world's food is produced on family farms, the vast majority of which is small to very small (Food and Agriculture Organization of the United Nations 2014). Small-scale women farmers represent the majority of rural poor populations in developing countries and their numbers are increasing (Rekha and Mary Hill 2008). The Gender in Agriculture Sourcebook points to the distribution of water and land as a major determinant of poverty, with women having far less access to these essential assets than men (The World Bank, FAO, IFAD 2008). Moreover, policy and decision-making regarding land and water management are traditionally in the male domain. As a result, policies and programmes do not always consider women's unique knowledge, specific needs or unequal ownership rights.

According to the International Fund for Agricultural Development (IFAD), the lack of attention for gender issues and women's unequal access to natural resources is the main obstacle to improving the livelihoods of the rural poor. Paying adequate attention to the water and gender interface also increases water efficiency and avoids expensive mistakes. In Sub-Saharan Africa for instance, gender-based farming systems with women and men cultivating separate fields are quite common. Ignoring this gender dimension has caused the partial or total failure of irrigation schemes. In the daily reality of rural women, the same water source typically meets multiple needs and there is no clear distinction between domestic and productive water. Not addressing the multiple uses of water has a proven negative effect on women's livelihood resilience, and has also been recognised as one of the causes of lower participation of women in Water User Associations (IFAD 2007).

The positive impact on women of investing in water for poverty alleviation and rural development is underscored by the World Bank OED report on focused Bank lending for rural water supply. This evaluation of 15 stand-alone water and sanitation projects reports, among other things, a dramatic decrease in time spent by women and children on water collection, in some cases of up to 80%, and a related increase in beneficial activities such as education, family hygiene and women's engagement in economic activities (Ronald and Tauno 2000).

Such examples illustrate that addressing water and gender equality together can create a positive multiplier effect. Gender equality and access to water can both have a catalytic effect on community development and they both can contribute positively to breaking the vicious cycle of poverty.

Box 7.1 Salamatu Garba, Director of Women Farmers Advancement Network of Nigeria (WOFAN):

Rural women in Nigeria are voiceless. Although they are the majority of the farmers, they have no say over water and no equal rights. Water management does not take into account the needs of those women; and sanitation – which is a major problem for women - is not high on the agenda. Also, we should realise that problems do not cut across and that capacities are very different. WOFAN therefore organises local women's groups and supports these groups to work in their own pace and context. We develop the capacity of local women to effectively address their problems. WOFAN has 1500 functional women groups. With support of USAID we have sunk over 1500 boreholes in four States of Northern Nigeria that are maintained by the women's groups themselves. In additional to water, WOFAN self-help groups address sanitation, climate resilient farming, marketing of products etc., which gradually transforms their lives. These women in rural Nigeria now speak out in public and address their issues with political leaders. They have become key actors in their own development.

Women Farmers Advancement Network of Nigeria (WOFAN) was established in 1993 and has grown to 1500 active working groups of 20–30 members in 5 states of Northern Nigeria

7.3 Gender Equality and Women's Empowerment: An Instrument for Development

Gender roles – and therewith the positions that men and women occupy in public and private life - may vary substantially between countries, cultures, ethnicities and generations. Traditional gender divides continue to restrain women's opportunities to lead productive lives. This makes gender equality not only a human rights principle but also an important contributor to societal development.

Since the 1970s important milestones such as the International Women's Year (1975), the United Nations Decade for Women: Equality, Development and Peace (1976–1985), and four World Conferences on Women,[1] have united the international community behind a common set of objectives to promote women's integration and full participation in development efforts, with concrete action plans to advance the status of women in public and private life. The efforts undertaken have gone through several phases and transformations, from regarding women almost exclusively in terms of their development needs, to recognizing their essential contributions to the entire development process, to seeking their empowerment and the promotion of their right to full participation at all levels of human activity (UN Women 1975–1995).

The development focus on women's equal participation continued into the twenty-first century with the Millennium Development Goals (MDG-3) targeting gender parity in education, enrolment in the labour market and increased political participation of women as instruments of women's social and economic empowerment (Millennium Development Goals 2015). However, in his report for the 60th Session of the Commission on the Status of Women (UN Economic and Social Council 2016), the United Nations Secretary General, concludes that: "Acknowledgement of the importance of women's social, economic and political empowerment by State and Non-state actors has not been matched by concrete policy implementation and demonstrable change on the ground".

The recent UN Women world survey on the role of women in development (UN Women 2014) underpins the interconnectedness between gender equality/women's empowerment and sustainable development. Women's agency is identified as central to many sustainable development pathways over the past 20 years in areas such as food security, climate change adaptation, and management of local ecologies including water, sanitation and energy provision, which are all critical elements for sustainable livelihoods for the poor.

The 2030 Agenda for Sustainable Development includes a dedicated goal for achieving gender equality and the empowerment of women and girls (SDG #5) and has many references to women as development agents.

[1] Mexico 1975, Copenhagen 1980, Nairobi 1985, Beijing 1995.

Box 7.2 Voices of Katosi Women Development Trust (KWDT)

Rehema Bavuma, a KWDT pioneer:

Water is a very strong empowerment tool. Everyone says that women suffer the most from absence of water. Yes, that is very true. But I have also seen how women can transform from mere water carriers into the best water managers. The women of Katosi Women Development Trust in Mukono District, Uganda, a fisher community on Northern shore of Lake Victoria, have organised themselves to address their water and sanitation challenges. Rainwater harvesting schemes now bring the water close to them and remove the burden of daily water collection; but equally important, this project has given the women confidence that they can change their lives and their communities. The Katosi women have lobbied their political leaders to ensure adequate water and sanitation in public places and they have thus developed sustainable agriculture programmes, generating income and diversifying the nutrition of their families. When you empower women, they can turn the world around.

Margaret Nakato, Director Katosi Women Development Trust:

Over the years KWDT has developed into a strong support structure for women in poor communities of Uganda. We engage in a long-term process: building grassroots groups, empowering individual women and increasing their participation in decision-making so that change comes from within. Access to safe water and adequate sanitation is our starting point. We have broken gender taboos by training women as masons so that they can build and maintain their rainwater harvesting tanks themselves. KWDT has instated a revolving fund from which community members can borrow to continuously acquire tanks and livestock so that communities continue to benefit.

Matovu Prossy of Bugolombe women's group:

Through my group, Bugolombe women's group, I applied for a tank from KWDT and one was constructed in our home. My husband was very proud of me and he has since treated me with respect. He does not see me as a dependent anymore but as someone who contributes to improvement of life.

(continued)

Box 7.2 (continued)

KWDT was formed in 1996 by 20 women from Mukono District in Central Uganda to improve the living standards in their community. Water and sanitation provision is a prime entry point. Today there are over 511 women organised into 19 groups. KWDT has so far been recognised nationally and internationally with several awards for transforming women's lives through the water and sanitation program among others. KWDT has also been selected by the Uganda Ministry for Water and Environment for the up scaling of rainwater harvesting through the revolving scheme in the whole district.

7.4 IWRM: A Participatory and Women Inclusive Approach

For the water domain, an integrated, participatory and women-inclusive approach is enshrined in the Dublin Principles for Integrated Water Resources Management (IWRM) of 1992 (The Dublin Statement on Water and Sustainable Development 1992).

IWRM is essentially a Human Rights-Based Approach (RBA). A key distinguishing feature of RBA is the process through which development outcomes are achieved. Normative principles that guide RBA development processes are: equality and non-discrimination, accountability, and participation. RBA transforms beneficiaries from passive recipients of development aid into active citizens with rights and expectations, but also with responsibilities. A fundamental benefit is that RBA unfolds the potential of the grassroots. Moving from receiving charity to claiming justice boost people's self-esteem. It liberates the underprivileged from their position as 'victims' and empowers them to actively pursue their rightful entitlements, to hold their governments accountable and to participate in decision-making on issues that concern them (Emilie Filmer-Wilson 2005).

There is mounting recognition that stakeholder participation in the development, implementation and management of water service provision is truly beneficial. In their comprehensive global assessment of water supply options, Sutherland and Fenn state to that effect (Sutherland and Fenn 2000): "*Whatever the circumstance (be it in rural Laos or urban California), the chances of improved performance and sustainability are greatly increased with the inclusion of all stakeholders in the planning and operation of water supplies.*" The Global Water Partnership describes participation of stakeholders across communities, civil society, minorities, indigenous peoples and women, and at all levels and stages, as an intrinsic element of effective water governance and conditional to achieving fair trade-offs between different water uses (Peter and Alan 2003).

Box 7.3 The Dublin Principles:
Dublin Principle 2: *Water development and management should be based on a participatory approach, involving users, planners and policy-makers at all levels. The participatory approach involves raising awareness of the importance of water among policy-makers and the general public. It means that decisions are taken at the lowest appropriate level, with full public consultation and involvement of users in the planning and implementation of water projects.*

Dublin Principle 3: *Women play a central part in the provision, management and safeguarding of water. This pivotal role of women as providers and users of water and guardians of the living environment has seldom been reflected in institutional arrangements for the development and management of water resources. Acceptance and implementation of this principle requires positive policies to address women's specific needs and to equip and empower women to participate at all levels in water resources programmes, including decision-making and implementation, in ways defined by them.*

Box 7.4 Voices of Women in Mweteni Village, Tanzania:

Resti Gerald, Chair of the Tegemeo Women Group:

Ten years ago water was a big problem for the 10 000 people of Mweteni village. Women and children had to spend a lot of time fetching water; the water was not clean and caused diseases; children missed school and when HIV Aids came into our village, the water burden became too much. When Tegemeo Women Group set out to address these problems, we were met with obstruction from our community and authorities who did not think women could do such a project. Together and with the help of Women for Water Partnership we have overcome all challenges and united the villagers and leadership to jointly develop and implement our plans with the Same District. Now clean water runs through pipes to all our sub-villages; our hospital and schools have rainwater harvesting systems and proper toilets; we have our water title deeds secured, water tariffs and water user associations that manage and maintain the systems. Tegemeo has also developed income-generating activities like sewing, planting maize and joint sale of products. And we have developed a revolving fund (vikoba) to finance new activities. Water has changed our lives.

Happy Zawadi of Mweteni:

This program empowered women; during meetings they are no longer silent, but we speak freely. There is no longer the problem of fetching water far away. Less female students drop out, they have time to do their homework.

Tegemeo Women Group (TWG) was formed in 2001 by seven women from Mweteni Village in Same District of Tanzania to promote gender equality and women's economic independence. TWG has taken a leading role in obtaining access to water for sustainable livelihood for the approximately 12 000 inhabitants of Mweteni and grown to 25 members in the process.

While women's involvement in the provision, management and safeguarding of water is generally embraced by development and donor agencies, an extensive literature review in 2007 considers the evidence base too fragmented to conclude that the positive effect of women's agency on water management is undisputed, nor that it yields unassailed benefits for the women themselves (Ray 2007). Local circumstances are very different and social roles not only vary across gender, meaning that women in different communities or segments of society may have different requirements for their empowerment. Enabling participation requires a good understanding of social conditions and bespoke approaches are needed that are best left to the local stakeholder groups themselves. The many best practices from women's civil society around the globe make a compelling case to further ascertain the conditions under which women can contribute meaningfully to water security and reap the benefits of their involvement.

7.5 The Role of Women's Civil Society

The remaining part of the global population that lacks access to water and sanitation is hard to reach. Many live in dispersed rural communities and informal settlements of developing countries, where water is often and increasingly scarce, traditional norms and gender divides prevail, and water allocation is guided by customary law in an informal water economy.

The earlier mentioned World Bank OED evaluation points out that the challenges for making decentralised rural water supply services work sustainably and reach all intended beneficiaries are typically of a social and institutional nature. Villages with higher social capital are reported to have much better results than villages where the level of social organisation was low. Social capital is what holds communities and groups together and guides their collective action. It is defined as: *the internal social and cultural coherence of society, the norms and values that govern interactions between people, and the institutions in which they are embedded.* Greater social capital in a community leads to more participation in service design, to more effective rules for governing implementation and to better monitoring of construction, use and maintenance.[2]

Women traditionally unite in social networks that are characterised by norms of trust and reciprocity. These quality social relations provide a social support and safety net and enable network members to collectively resolve problems while obtaining mutual benefits. Women's social networks are organised at local, national and international level and in a diversity of peer groups ranging from professional background, religious or political affiliation, ethnicity or nationality, to thematic interest groups. They form a substantial resource for collective action contributing to social cohesion, democracy and sustainable development (Bouman-Dentener and Devos 2015).

[2] Pages 31–32 (Ronald and Tauno 2000).

A recent publication on women's agency in water illustrates how water and sanitation interventions in which women's civil society groups had a leading role not only effectively meet women's practical water needs and increase their productive contributions, but also facilitate women's engagement in public life, giving way to more inclusive decision-making, demand-responsive services, and increased cooperation with local authorities and the water sector. The positive effects of empowering women through water extend far beyond the water domain, as these grassroots groups subsequently address other development challenges in their communities (Bouman-Dentener 2015).

The United Nations University Institute for Water, Environment and Health (UNU-INWEH) has conducted 35 in-depth local level surveys to gain insight in what makes women's projects successful, how women's social capital adds to the sustainability of local WaSH interventions and what are the benefits of a supporting network at different scales. The projects typically are relatively small and civil society initiated, have multi-stakeholder participation and practice learning by doing. They include indigenous knowledge and leverage community contributions bringing about significant change with small financial investments. Considerable time and effort are invested in information sharing, capacity development, social dialogue and building trust and understanding, whereby technically viable solutions become accepted by the community and local ownership is built. The main challenges are recognition by and cooperation with authorities, gender stereotyping and inclusion of marginalised voice; these challenges are overcome through peer support and perseverance, and if need be, using the power of the global network. A main sustainability factor is that prior to project implementation, the absorption capacity of the community is built. Women's civil society with its intrinsic social capital and bridging and bonding capacity proves an effective and low-budget medium to ensure community involvement, acceptance and trust. If this process precedes the actual project development phase, true demand driven and locally owned processes result. The global network serves to build broader support for common values and goals, to build knowledge hubs and partnerships, to share experiences, to give local and national women's groups recognition, to provide support in fundraising and capacity development, and to influence (inter)national agenda setting (Schuster-Wallace et al. 2015).

7.6 Bridging the Remaining Global and Gender Divides

On the brink of the 2030 development era just under three quarters of a billion people are reported to not have access to improved drinking water sources and 2.5 billion people lack access to basic sanitation, with over one billion people still defecating in the open (WHO/UNICEF 2014). Indigenous peoples, minorities and the rural poor are predominant segments of the global population that still lack adequate access to safe and affordable water for their daily needs. They are also the ones that often experience gender inequalities in rights, resources and voice. This makes women's empowerment and access to water important and complementary development instruments. Working on the SDG#5 – SDG#6 interface tackles poverty at its roots.

Closing the remaining global and gender divides in water means getting to the hard-to-reach population segments that often live in conditions where centralised water management systems are not an option. Decentralised and tailor-made solutions are needed that take the specific physical circumstances and socio-cultural contexts into consideration. Decentralised water supply and management works best when there is sufficient social capital to engage communities and guide their collective action. Women have a tradition of working through social networks. Making use of women's civil society to mobilise their peers and their communities helps to prevent expensive mistakes and to make the complex network of local actors and social interactions in rural water provision work effectively, which, according to the cited World Bank evaluation,[3] is invariably more elaborate and time-consuming than anticipated. Women's civil society knows the social norms and customary laws and has the proximity needed to render moral support for a prolonged period of time.

Building social capital through empowering grassroots women's groups requires effort, time and resources, but such pre-investment would contribute to sustainable water management and community development as a whole. While international development policies have put gender equality and women's empowerment front and centre on the development agenda, this strategic focus is thus far not reflected in gender responsive budgeting. OECD-DAC[4] uses a gender equality policy marker to assess the gender focus of Official Development Aid (ODA). An analysis of overall donor spending in DAC countries in 2012 reveals that of the 86 061 million USD that were screened for the gender equality policy marker (87.1% of the total allocated amount), 27% included gender equality/women's empowerment as an explicit or as a secondary objective of the activity. A mere 2% of this so-called gender-focused funding was allocated to women's equality organisations and institutions (OECD 2014).

Poor funding marginalises the role of women's civil society and restricts their contributions to gender sensitive water management. At the same time, audits and evaluations by, inter alia, the European Union (European Court of Auditors 2012), the Dutch Ministry of Foreign Affairs (Evaluation 2012) and the World Bank (1997–2007) indicate that a considerable number of water and sanitation projects fails to adequately service the targeted beneficiaries. Meaning that many millions invested in water and sanitation provision are effectively wasted on unsustainable interventions partly because, due to insufficient involvement of the beneficiaries, interventions fail to match their demands.

[3] Page xii (Ronald and Tauno 2000).

[4] Organisation for Economic Cooperation and Development – Development Assistance Committee.

The 2030 Agenda for Sustainable Development calls for an all-of-society engagement and partnership, clearly articulating the importance of involving non-State actors including civil society, in the implementation of this ambitious agenda. It raises the expectation that important lessons have been learned from the past and that world leaders are willing to seriously pre-invest in strengthening those stakeholders that are often and easily forgotten, but known to be crucial for the transition from principles to practice. Women are such a stakeholder group.

Going from the conceptual comprehension of gender issues to solving the everyday grassroots realities of differential access to and use of water is a long and winding road. If we are serious about the 2030 Agenda for Sustainable Development, now is the time to translate the rhetoric of women's important role in the provision, management and safeguarding of water into action and seriously pre-invest in the empowerment of women and other easily forgotten stakeholder groups to contribute meaningfully to the SDG#5 –SDG#6 interface.

References

Bouman-Dentener, A. M. (2015). *Women as agents of change in water: Reflections on experiences from the field.* A WfWP publication in cooperation with UN Women and UNW-DPAC. Available from: http://www.womenforwater.org

Bouman-Dentener, A. M., & Devos B. (2015). *Civil society: Key contributors to water and sustainable development.* UN Water annual international Zaragoza conference "Water and Sustainable Development: from Vision to Action". Available from: http://www.un.org/waterforlifedecade/waterandsustainabledevelopment2015/pdf/OP_CivilSociety_4themes_FORMAT.pdf

Emilie Filmer-Wilson. (2005). *The human rights-based approach to development: The right to water* (Netherlands quarterly of human rights, Vol. 23/2). Available from: http://www.crin.org/docs/right-to-water.pdf

European Court of Auditors. (2012). *European union development assistance for drinking water supply and basic sanitation in Sub-Saharan Countries.* Special report No. 13. http://www.eca.europa.eu/Lists/ECADocuments/SR12_13/SR12_13_EN.PDF

FAO. (2015). *AQUASTAT website, Food and Agriculture Organization of the United Nations (FAO).* Website Accessed 25 Jan 2015. Available from: http://www.fao.org/nr/water/aquastat/water_use/index.stm

Food and Agriculture Organization of the United Nations. (2014). *International year of family farming: Legacy of IYFF and the way forward.* Available from: http://www.fao.org/3/b-mm296e.pdf

General Assembly Resolution 70/1. *2030 Transforming our world: The agenda for sustainable development,* (21 October 2015). Available from: http://www.un.org/ga/search/view_doc.asp?symbol=A/RES/70/1&Lang=E

IFAD. (2007). *Gender and water: Securing water for improved rural livelihoods, the multiple-uses system approach.* Rome: International Fund for Agricultural Development. Available from: http://www.ifad.org/gender/thematic/water/gender_water.pdf

IOB Evaluation. (2012). *From infrastructure to sustainable impact: Policy review of the Dutch contribution to drinking water and sanitation (1990–2011).* Dutch Ministry of Foreign Affairs. Available from: http://www.oecd.org/derec/netherlands/IOB%20nr%20%20366%20From%20infrastructure%20to%20sustainable%20impact.pdf

Millennium Development Goals. (2015). Available from: http://www.un.org/millenniumgoals/gender.shtml

OECD. (2014). *Creditor reporting system database*. Aid in support of gender equality and women's empowerment – Donor charts. Statistics based on DAC members' reporting on the Gender Equality Policy Marker, April 2014. Available from: http://www.oecd.org/dac/stats/documentupload/Aid-support-gender.pdf

Peter, R., & Alan, H. (2003). *Effective water governance*. GWP-TEC background paper #7. Available from: http://www.gwp.org/en/ToolBox/PUBLICATIONS/Background-papers/

Ray, I. (2007). *Women, water and development. Annual Review of Environment and Resources, 32*, 421–449 Available from: http://www.annualreviews.org/doi/pdf/10.1146/annurev.energy.32.041806.143704.

Rekha, M., & Mary Hill, R. (2008). *A significant shift: Women, food security and agriculture in a global market place*. International Center for Research on Women (ICRW). Washington, DC. Available from: http://www.icrw.org/sites/default/files/publications/A-Significant-Shift-Women-Food%20Security-and-Agriculture-in-a-Global-Marketplace.pdf

Ronald, P., & Tauno, S. (2000). *Rural Water Projects: Lessons from OED Evaluations*. OED working paper series no. 3. The World Bank, Washington D.C. Available from: http://www-wds.worldbank.org/external/default/WDSContentServer/WDSP/IB/2000/11/04/000094946_00102111465146/Rendered/PDF/multi_page.pdf

Sadoff, C. W., Hall, J. W., Grey, D., Aerts, J. C. J. H., Ait-Kadi, M., Brown, C., Cox, A., Dadson, S., Garrick, D., Kelman, J., McCornick, P., Ringler, C., Rosegrant, M., Whittington, D., & Wiberg, D. (2015), *Securing water, sustaining growth: Report of the GWP/OECD task force on water security and sustainable growth* (p. 22). Oxford: University of Oxford. Available from: http://www.gwp.org/Global/About%20GWP/Publications/The%20Global%20Dialogue/SECURING%20WATER%20SUSTAINING%20GROWTH.PDF

Schuster-Wallace, C. J., Cave K., Bouman-Dentener A. M., & Holle F. (2015). *Women, WaSH, and the water for life decade*. United Nations University Institute for Water, Environment and Health and the Women for Water Partnership. Available from: http://inweh.unu.edu/women-wash-water-life-decade/

Sutherland, D. C., & Fenn, C. R. (2000). *Assessment of water supply options*, Thematic Review IV.3 prepared as an input to the World Commission on Dams, Cape Town. Available from: http://www.dams.org

The Dublin Statement on Water and Sustainable Development. (1992). Available from: https://www.wmo.int/pages/prog/hwrp/documents/english/icwedece.html

The Millennium Development Goals Report. (2014). United Nations, New York. Available from: http://www.us.undp.org/content/dam/undp/library/MDG/english/UNDP_MDGReport_EN_2014Final1.pdf

The World Bank, FAO, IFAD. (2008). Gender in Agriculture Sourcebook. Module 6: *Gender Mainstreaming in Agricultural Water Management*. Available from: http://www.ifad.org/gender/pub/sourcebook/gal.pdf

UN DESA. (2010). The World's Women 2010: Trends and Statistics. Chapter 7 – Environment, pp. 142-145. ST/ESA/STAT/SER.K/19. Available from: http://unstats.un.org/unsd/demographic/products/Worldswomen/WW_full%20report_BW.pdf

UN Economic and Social Council. (2016). Women's empowerment and the links to sustainable development; report of the Secretary General. E/CN.6/2016/3. Available from: http://www.un.org/ga/search/view_doc.asp?symbol=E/CN.6/2016/3&referer=http://www.unwomen.org/en/csw/csw60-2016/official-documents&Lang=E

UN Women. (2014). *The world survey on the Role of Women in Development: Gender Equality and Sustainable Development*. New York: United Nations Available from: http://www.unwomen.org/en/digital-library/publications/2014/10/world-survey-2014.

UN Women – The Four Global Women's Conferences 1975–1995: Historical Perspective. Available from: http://www.un.org/womenwatch/daw/followup/session/presskit/hist.htm

WHO/UNICEF. (2014). *Progress on drinking water and sanitation, 2014 update*. Geneva: World Health Organisation Available from: http://www.unicef.org/gambia/Progress_on_drinking_water_and_sanitation_2014_update.pdf.

World Bank Independent Evaluation Group. (2010). An Evaluation of World Bank Support, 1997–2007. Volume 1: Water and Development. Available from: http://ieg.worldbank.org/Data/reports/water_eval_0.pdf

World Water Assessment Programme. (2013). UN WWAP UNESCO project for gender sensitive water monitoring, assessment and reporting. Available from: http://www.unesco.org/new/en/natural-sciences/environment/water/wwap/water-and-gender/

WWAP (United Nations World Water Assessment Programme). (2015). *The United Nations World Water Development Report 2015: Water for a Sustainable World* (p. 19) . Paris: UNESCO. Available from: https://www.unesco-ihe.org/sites/default/files/wwdr_2015.pdf

Chapter 8
Public Health Dimensions of Water Insecurity

Susan Watt

Abstract Water insecurity poses challenges to many sectors such as energy, housing, agriculture, and health. Similarly, public health addresses the biological, social, and psychological determinants of health at the population level and consequently must attend to issues in the same multiple sectors. This chapter examines the relationship between water insecurity and public health. It is argued that, in their primary roles of disease prevention/protection, mitigation, adaptation and health promotion, public health providers must lead initiatives to improve quality and ensure adequate quantities of water in order to sustain livelihoods, human well-being, and socio-economic development. Reciprocally, public health should be included in the development of public policies and community planning for water resources. Ten intersecting areas of water and public health concern are discussed: acute infection; chronic infection; food safety and security; malnutrition; maternal and newborn health; environmental integrity; disaster management; population growth; population safety; and, health information dissemination. Researchers in both development and public health need to increase their attention to identifying and evaluating the impacts of water as a hazard and of risk management strategies that can prevent, ameliorate, or mitigate water insecurities. Water security analysts need to include public health considerations in their analysis of security risks. Finally, the chapter provides examples of the intersection of water insecurity and public health from both industrial and developing parts of the world.

Keywords Public health • WaSH • Water insecurity • Disease • Policy

S. Watt (✉)
School of Social Work, McMaster University, Hamilton, ON, Canada

United Nations University Institute for Water, Environment and Health,
Hamilton, ON, Canada
e-mail: wattms@mcmaster.ca

D. Devlaeminck et al. (eds.), *The Human Face of Water Security*, Water Security in a New World, DOI 10.1007/978-3-319-50161-1_8

8.1 Water Insecurity and Public Health – An Historic Perspective

Public health is "the science of protecting and improving the health of families and communities through promotion of healthy lifestyles, research for disease and injury prevention and detection and control of infectious diseases."

Source: http://cdcfoundation.org/content/what-public-health, 4 Sept 2014.

Public health, in its most rudimentary form, is the taking of collective responsibility (and action) to improve health and to prevent disease and disability in the population. The roles of public health providers include disease prevention/protection, disease mitigation and adaptation, and health promotion. At different times and in diverse circumstances, there is a variable emphasis on each of these aspects of public health. When the right approach is used at the right time, under the right circumstances, the net result is to decrease morbidity and mortality and to improve the health status of targeted groups. Public health is concerned with population health, with the well-being of groups of people, rather than with an individual's health and illness. While the impact of public health is felt at the individual level (e.g. immunization of individuals), the public health focus is on community at all levels. Therefore, for the purposes of this work we will focus on communities, from the village through to international considerations, leaving the important discussion about individuals (e.g. the ethics of individual vs. community responsibility for health) to other authors (Fig. 8.1).

The initial focus of modern public health was on protection – on disease prevention and especially on the prevention of large scale outbreaks of contagious diseases (Rosen 1993). Indeed, the history of public health is dominated in the literature by a Eurocentric and North American perspective, relying heavily on descriptions of

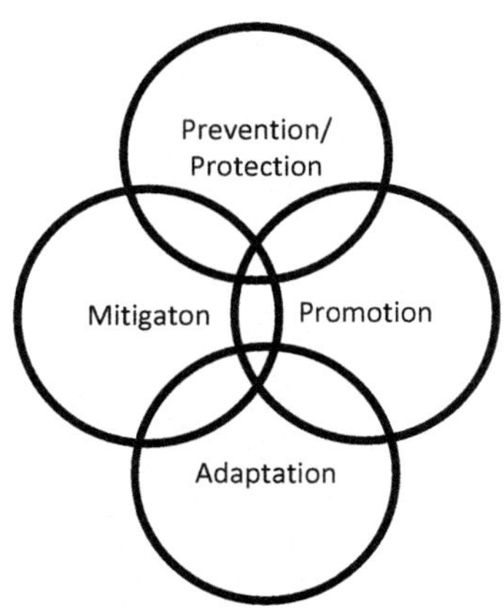

Fig. 8.1 The role of public health

the administrative structures of public health agencies and paralleling the history of medicine (Fee 1993).

However, archeologists and anthropologists have uncovered global evidence that the earliest civilizations paid attention to community health, and especially to the provision of water and community sanitation structures. Similarly, it is documented that all known societies, from their origins, have paid attention to cleanliness of both the individual and the environment with rules about the washing of human bodies, the preparation of food, and the burial of the dead (Allegranzi and Pittet 2009; Curtis et al. 2009). These rules are often encoded in religious tomes that articulate the use of water for initiation, healing, restoration, and purification (Chamberlain 2007).

Public health's original focus on water was based on an observational understanding. Seasons and locations were seen to result in water insecurity from both too much and too little access to water and consequent disease outbreaks (Cowell 2001; Winter 2013). Water security was a determining factor in locating communities for access not only to water for drinking, cleaning, and agriculture but also to avoid miasmas (Anderton and Leonard 2004) and to support for transportation and industrial production (Hall 2014). Scientific knowledge enhanced our observational understanding and led us to a biological understanding of why clean water, in sufficient amounts to meet human and agricultural needs, is essential for survival but also led to a refocusing on protection from disease. The sanitation movement, a product of industrialization and urbanization in the late nineteenth century in North America and Europe, placed clean water and sanitation once again on the public health agenda (Beaglehole and Bonita 1997).

In the twentieth century, the development of our rudimentary understanding of bacteriology and virology led to increased attention to water quality as a factor in disease prevention as the concepts of waterborne diseases and water as a disease vector took hold. Prevention and remediation strategies included purification, sterilization, and vaccination in attempts to control disease spread, and the use of antibiotics in the treatment of disease. Secondary prevention in the form of water treatment was assumed in developed countries and little attention was paid to either source purity or sufficiency of water supplies. In industrialized countries, public health emphasis was placed on regulating and controlling water for commercial use and public consumption. Urban planning was dominated by public health knowledge and led to the development of cities with large suburban tracts that separated work and living environments (Sarkar et al. 2014). It was generally held that, although there could be problems with water-related diseases, these problems could be solved by pharmaceuticals and prevented by regulations. Progress was dominantly measured in economic terms with little attention to environmental considerations. In the public health sector modest attention was paid to the interrelationship between water and public health, but access to potable water became an increasingly important part of the development agenda (Beaglehole and Bonita 2008).

At the beginning of the twenty-first century global attention has begun to address environmental issues framed largely by the climate change debates (Bouzid et al. 2013; McMichael 2013). Water related disasters have resulted in outbreaks of water-related diseases such as cholera and typhoid and lack of water in non-water related disasters have led to similar outbreaks (Alexander et al. 2013; Ebi et al. 2013). As globalization of the economy has increased awareness of health problems in low and middle income countries (LMICs)[1] and the vulnerability of travellers from high income countries (HICs), and the cost of treating diseases has soared, international public health practitioners and policy makers have turned their attention to the promotion of health. This refocusing led to the discovery of the importance of poor water, inadequate sanitation and lack of personal hygiene as factors in global mortality, morbidity, and disability. In fact, 6–8% of disability adjusted life years (DALYs) were attributed to these factors (Murray and Lopez 1997). It is not surprising that the most vulnerable, the sick, the elderly, and children, are most at risk of dying from these infections. In this context, consistent, sufficient, usable water has once again emerged as a public health concern (Desai 2013; Mulligan et al. 2012; Azizullah et al. 2011). Additionally, water insecurity as a result of climate change has emerged as a concern for disaster management (Smith 2013; Scott et al. 2013; Habiba et al. 2013; Field 2012).

At the same time, conceptually, public health began to embrace the paradigm of iterative involvement in planning, implementation, and evaluation. This model, which is similar to disaster response stages and climate change response holds the potential for direct application in in the water sector (Fig. 8.2).

8.2 Water Insecurity

Water security is defined as the capacity of a population to safeguard sustainable access to adequate quantities of acceptable quality water for sustaining livelihoods, human well-being, and socio-economic development, for ensuring protection against water-borne pollution and water-related disasters, and for preserving ecosystems in a climate of peace and political stability (UN-Water 2013).

For much of the world, water insecurity is the norm. The water supply to many communities is neither sufficient nor safe for human consumption; one in six people worldwide do not have access to improved water sources (World Health Organization 2013a; UNICEF, World Health Organization 2012). Researchers

[1] Annually, LMICs and HICs are defined by the World Bank from the analytical classification of the world's economies based on estimates of gross national income (GNI) per capita for the previous year. As of July 2015, low-income economies are defined as those with a GNI per capita, calculated using the World Bank Atlas method, of $1045 or less in 2013; middle-income economies are those with a GNI per capita of more than $1045 but less than $12,746; high-income economies are those with a GNI per capita of $12,746 or more. Lower-middle-income and upper-middle-income economies are separated at a GNI per capita of $4125. [From http://data.worldbank.org/news/2015-country-classifications , retrieved August 9, 2015]

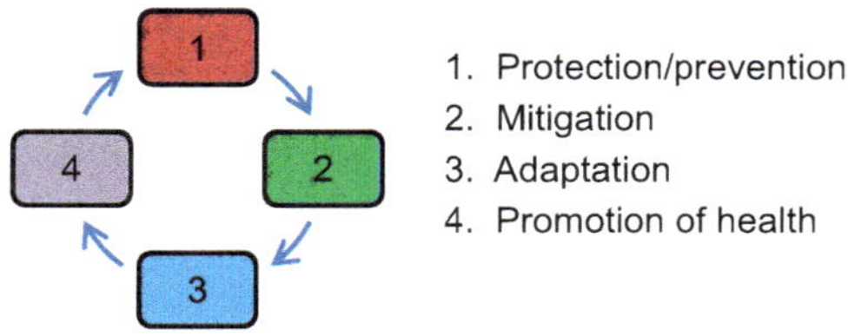

Fig. 8.2 The cycle of public health intervention

focused on climate change predict that this situation will worsen as the impacts of climate change extend to cover larger geographic areas and have greater impacts on water supplies (Allan et al. 2013). Concurrently, when agriculture commands 70% of fresh waste use, food insecurity becomes another concern for public health as will the increased stress placed on water by hydroelectric and biofuel demands, population growth (WWDR 2012) and migration from rural to urban settings (World Health Organization 2006). The Intergovernmental Panel on Climate Change (IPCC) points towards a great vulnerability of freshwater resources as a result of climate change, with severe consequences for economic, social and ecological systems (IPCC 2008, 2012).

Lobell et al. predict that South Asia and Southern Africa will be the most vulnerable to both food and water climate change related shortages and will be affected by factor such as human security, economies, political dealings, and institutional capacity to generate, sustain, and deliver adequate amounts of safe water (Lobell 2014).

> Addressing water security, therefore, requires interdisciplinary collaboration across sectors, communities and political borders, so that the potential for competition or conflicts over water resources, between sectors and between water users or states, is adequately managed (UN Water, p. 12).

Water insecurity threatens the health and physical safety of communities around the world. Both people and wildlife depend on a consistent and adequate supply of safe water to sustain life; both migrate to water sources (Black et al. 2011; Dingle 1996). When water is unsafe to drink, too scarce as in the case of drought, too plentiful as in the case of floods, or unpredictable as is the case when climate change impacts weather patterns and leads to an insecure food supply, both humans and animals are at risk. Although migration is driven by many complex factors, seeking potable water is a major driver of both mass and rural-to-urban migration. Water extremes – drought and flooding – jeopardize physical security from a lack of food, water, and shelter. Migration frequently leads to secondary problems in food, water, and shelter exacerbated by serious economic disruption and disease outbreaks (Black et al. 2013; Basu and Shaw 2013; Sternberg 2014). How to prevent, mitigate, or adapt to water insecurities and how to promote healthy environments and behaviours under these circumstances defines the intersection of public health and water insecurity.

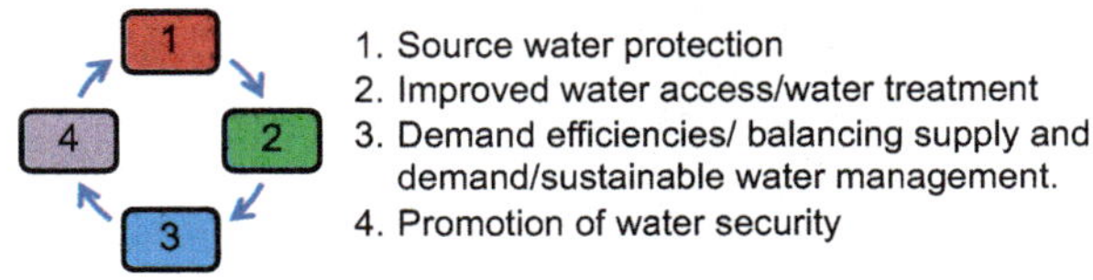

Fig. 8.3 The cycle of water security

8.3 Water and Public Health: A Framework for Engagement

Using the public health iterative framework, water security can be viewed as the same cyclical action. These activities can be in the form of planning, policy, research and practice (Fig. 8.3).

The public health tools that can be used to approach water insecurity include surveillance (measurement), impact assessment (HIA) and environmental planning. Surveillance establishes the basic characteristic of water availability – when, where, how much, of what quality water. For any planning to be done surveillance is an essential but often overlooked step. It takes time, money, and careful attention to both overall patterns and details of all water sources and their accessibility over time. Health surveillance requires a similarly comprehensive approach.

> Health Impact Assessment (HIA) is a means of assessing the health impacts of policies, plans and projects in diverse economic sectors using quantitative, qualitative and participatory techniques.
>
> Source: http://www.who.int/hia/en/. Accessed November, 2014.

An impact assessment is used to maximise the positive and minimise the negative impacts of policies and activities in a specific community based on existing knowledge and experimentation. It can be done in advance of a change in policy or practice or concurrently with the new activity. For example, WHO and UNICEF through their joint monitoring programme have undertaken massive data collection to determine the impact of MDGs on access to water and sanitation and how this is reflected in the health status of nations (Bradley and Bartram 2013). As a result, in addition to the information on specific initiatives (i.e. water and sanitation provision projects) undertaken by private, public, and civil society organizations, information is available on the impact of these initiatives on their communities and on the national health status of the countries in which the communities are located. Similarly, Cheng et al. provide evidence of the impact of water and sanitation access on maternal and child health (Cheng et al. 2012).

On the water security side, the Soil Water Assessment Tool (SWAT) provides a way of modeling watershed properties, water quality, and watershed management (Tuppad et al. 2011) thus providing public health with information about impending water insecurity. Environmental impact assessment tools provide information on the impact of built environments. This information is important to public health practitioners and policy makers in relation to both water systems management and disaster planning and remediation, especially in urban areas.

Environmental planning is the third public health tool used in relation to water insecurity. It is the process of understanding and making decisions about the relationships between human and natural systems. From a public health perspective, environmental planning that focuses on water security for both the individual (i.e., having consistent access to safe water in quantities to meet personal needs) and the community (i.e., preventing flooding and providing water reserves for dry periods) results in multiple impacts. In both cases source waters need to be protected from contamination and require equitable allocation between competing interests and require equitable allocation between competing interests (Jackson et al. 2012). Failure to prevent water insecurity and to repair breeches in water security results in both acute and chronic diseases and secondary impacts such as jeopardising food security and maternal and newborn health (Vörösmarty et al. 2010). Therefore, environmental planning requires collaboration and cooperation between planners and public health practitioners at both the policy and the practice levels.

The developing world is replete with planning initiatives that failed to take public health into account. A well that is dry because it was not properly sited is an engineering problem, but a well that is not used because it is in a dangerous location for the women who haul water is a public health issue. Both contribute to water insecurity and to public health insecurity. Building a health care facility without safe water and sanitation resources is antithetical to public health principles. Locating a latrine above a water supply in the backyard of a school on a hillside jeopardizes the health of all the children in the school. Technology in developing countries is not the problem; but from a public health perspective, its application may be.

Public health has many responsibilities, often encoded in legislation, in high income countries. In LMICs these functions often are the domain of NGOs and international donor groups. It is helpful to identify the most common services that fall within the scope of practice of public health.

Why then should water insecurity be a focus of public health concern?

Box 8.1 Ten Essential Public Health Services

1. Monitor health status to identify community health problems.
2. Diagnose and investigate health problems and health hazards in the community.
3. Inform, educate, and empower people about health issues.
4. Mobilize community partnerships to identify and solve health problems.
5. Develop policies and plans that support individual and community health efforts.
6. Enforce laws and regulations that protect health and ensure safety.
7. Link people to needed personal health services and assure the provision of health care when otherwise unavailable.
8. Assure a competent public health and personal health care workforce.
9. Evaluate effectiveness, accessibility, and quality of personal and population-based health services.
10. Research for new insights and innovative solutions to health problems.

(continued)

Box 8.1 (continued)

The Community Tool Box is a service of the Work Group for Community Health and Development at the University of Kansas

Source:http://ctb.ku.edu/en/table-of-contents/overview/models-for-community-health-and-development/ten-essential-public-health-services/main. Accessed 4 Sept, 2014.

8.4 Water as an Essential Health Requirement

Water is a multi-dimensional issue and a prerequisite for achieving human security, from the individual to the international level (UN-WATER 2013).

Although there are a few epidemiological studies that demonstrate the association between insecure water supply and poor health linking outbreaks to municipal water supplies (Davis et al. 2014), the association is clearly established from a clinical perspective. According to the Mayo Clinic, the average healthy human requires between 9 and 13 cups (2.2–3.1) of water each day in a moderate climate. Increased temperature, exertion, or illness may increase that need.

In addition, individuals require uninterrupted access to sufficient amounts of safe water for sanitation and hygiene practices and for food preparation. Water that is insufficient, unpredictable, or contaminated compromises health from infection (Gundry et al. 2003) and from dehydration (Malm et al. 2013). During the recent Ebola crisis in West Africa it was estimated that each patient requires 150 litres of water per day to meet their personal WaSH needs as opposed to the usual WHO recommendation of 20 l/day.

8.5 Water Security as a Public Health Issue

The water community has a need for a range of services to support decisions related to planning including:

- Identification of extreme weather and climate hazards that pose water-related risks;
- Identification of populations vulnerable to weather and climate hazards, including those in the coastal zone;
- Allocation and re-allocation of water resources;
- Design and placement of infrastructure and personnel (i.e. water management organizations, structures and facilities);
- Implementation of risk management and emergency preparedness practices and procedures;
- Dissemination of information to users, including the public, i.e. Public Service forecasts and alerts);
- Development and implementation of water and environmental policy;
- Development and implementation of water and flood management policies and strategies;
- Development and implementation of water management regulations and laws

Source: http://www.ncbi.nlm.nih.gov/pmc/articles/PMC2819784/#b24-ijerph-07-00189. Accessed 27 October, 2014.

Investing in water provides benefits at many levels and across many sectors. The benefits include health, quality of life, economic prosperity, and environmental integrity (World Health Organization 2014a). Indeed there are ten specific areas in which public health and water (in)security intersect and interact. Each area will be discussed.

Box 8.2 Public Health Foci on Water Insecurity

1. Acute infection
2. Chronic infection
3. Food safety and security
4. Malnutrition
5. Maternal and newborn health
6. Environmental integrity
7. Disaster management
8. Population growth
9. Population safety during armed conflict
10. Health information dissemination

8.5.1 Acute Infectious Disease

The expertise that a public health perspective adds to understanding water insecurity and its impact on the human condition covers a spectrum of concerns. Nowhere is the link between public health and water insecurity clearer than in relation to waterborne infections, water-washed infections, water-based infections, and infections with water-related insect vectors diseases (Moe et al. 2007; World Health Organization 2013b). Although the relationship between water and disease is known at a microbiologic level, little research has been done to provide a global picture of waterborne diseases and the factors which influence global patterns of prevalence and severity (Yang et al. 2012).

Until now these diseases have been a greater burden in warm climate regions, but all countries report some type and level of waterborne disease (Nichols et al. 2014). When water is chronically contaminated or when acute contamination occurs as a result of natural disaster or failed water purification systems, these infections increase (Kouadio et al. 2012). A recent example of this outcome is found in Haiti. Following an earthquake in 2010 that destroyed its already fragile water and sanitation structure, cholera emerged as the major public health concern. As of November 2014 the UN reported 707,000 suspected cases and over 8600 deaths from cholera in Haiti. The source cause of the cholera outbreak has been debated but the role of water in its spread is undisputed (Fung et al. 2013). Water insecurity as a direct result of contamination by sewage has been identified as the cause of this outbreak (Rojias 2014).

This public health disaster required a two-pronged approach: treatment of those already infected with cholera and prevention of its spread. In this case, water insecurity directly threatened public health. Providing a stable and secure source of potable water and securing water for sanitation and hygiene became a major public health initiative augmented by cholera vaccination.

8.5.2 Chronic Disease

The line between acute and chronic disease is often blurred in relation to water insecurity. Diarrhea, most often caused by waterborne pathogens, in its acute form can cause severe illness leading to death from dehydration and multiple systems failure. Worldwide, waterborne diarrhoeal diseases each year kill an estimated 1.8 million people of which 88% is attributable to the environment, including risks associated with unsafe water, lack of sanitation and poor hygiene (Braks and De Roda Husman 2013). Chronic diarrhea accounts for significant death rates in children less than 5 years of age; globally, it is the second leading cause of death in this age group. Chronic diarrhea reduces productivity in the educational, agricultural, domestic, and commercial sectors and often leads to malnutrition from malabsorption. It also places higher demands on water resources for hydration, bathing, medication. Additionally, chronic parasitic conditions such as schistosomiasis, caused by water which harbours worms of the schistosoma type found in Africa, Asia, and South America and affecting almost 210 million people worldwide, is an example of a chronic condition leading to disability and death (Fenwick 2012; King 2010). In the case of schistosomiasis, infection has been shown to be linked to the development of bladder cancer (Honeycutt et al. 2014).

Malnutrition is also a chronic condition associated with water insecurity. Severe conditions associated with malnutrition include marasmus (chronic wasting); cretinism and irreversible brain damage due to iodine deficiency; and blindness from vitamin A deficiency. The inability to work as a result of this illness has a major economic impact on the ability of both the individual and the community to produce adequate food supplies, thus drawing food security into the mix of factors related to public health and water (in)security.

Both acute and chronic health impacts have been experienced as a result of heavy metal contamination of drinking water. Specifically, arsenicosis, the result of high levels of arsenic in drinking water, is a chronic health problem that afflicts tens of millions across the world and has been found to lead to other conditions such as cancers of skin, liver, lungs and kidneys. The WHO has determined that high levels of ground water contamination from arsenic can be found in Argentina, Bangladesh, Chile, China, India, Mexico, and the United States of America. Drinking-water, crops irrigated with contaminated water, and food prepared with contaminated water are the most common sources of exposure (Gillispie et al. 2015).

Other heavy metals (chromium from leather tanning, mercury for industrial activities, for example) have also led to significant chronic, but less well studied health impacts. Additionally, aluminium, beryllium, cadmium, copper, iron, lead, lithium, manganese, thallium, tin, and zinc have been identified as toxic to humans when consumed in sufficient quantities either in a single episode or through chronic ingestion and resultant bioaccumulation. Human exposure is most often related to contamination of drinking water, livestock, or crops that are subsequently ingested by humans. Although these metals occur naturally their toxicity is related to the concentration of their consumption and to their solubility. Water is a common medium for the dissolution of the metal oxide and as a subsequent conveyor into the systems of humans, animals, and crops (Naggar et al. 2014).

8.5.3 Food Safety and Security

Water insecurity, in all its forms, jeopardizes food safety and security by producing food scarcity and poor food quality. Both flooding and droughts compromise food growing and storage conditions. Irrigation with untreated wastewater threatens the integrity of the foods grown under these conditions (Ashraf et al. 2013), but without this water source, jeopardizes crop production. On a small scale, local producers often cannot meet local demands and households cannot prepare foods with safe and reliable water. On a national and international scale food cannot be grown or prepared for exported, thus compromising food chain supplies to dependent populations at home and abroad (Hanjra et al. 2013).

In many parts of the world food safety is highly regulated by national and local governments. These regulations are enforced through formal public health mechanisms such as food and restaurant inspectors. While standards vary from place to place, the overall goal is to ensure that food is safe for human consumption. When food is transported across international borders, customs authorities include food importation restriction and safety inspections as part of their responsibilities.

8.5.4 Malnutrition

Malnutrition is both a cause and effect of water insecurity. Water insecurity contributes to malnutrition though two paths – diarrheal diseases from water borne pathogens and food insecurity secondary to water insecurity. There are complex links between food, water, poverty, and malnutrition (Headey 2013) that place children and the elderly at special risk for acute and chronic disease and disability (Ngure et al. 2014). In short, water insecurity jeopardizes food supplies leading directly to malnutrition. Malnutrition compromises the ability of a person to seek out and collect water. Increasing distances to water sources coupled with unreliable quantity and quality place extra caloric demands on already deficient water collectors who are largely women and children.

Box 8.3 Facts About Worldwide Malnutrition

1. 795 million people in the world do not have enough food to lead a healthy active life (one in nine);
2. The vast majority of the world's hungry people live in developing countries;
3. Asia is the continent with the most hungry people – two thirds of the total;
4. Sub-Saharan Africa is the region with the highest *prevalence* (percentage of population) of hunger (one in four);
5. Poor nutrition causes nearly half (45%) of deaths in children under five (3.1 million children each year);
6. One out of six children in developing countries is underweight;
7. One in four of the world's children are stunted;
8. If women farmers had the same access to resources as men, the number of hungry in the world could be reduced by up to 150 million;
9. 66 million primary school-age children attend classes hungry across the developing world (23 million in Africa alone); and,
10. The World Food Program calculates that US$3.2 billion is needed per year to reach all 66 million hungry school-age children.

Source: http://www.wfp.org/hunger/stats, 9 August, 2015.

8.5.5 *Maternal and Newborn Health*

Water insecurity impacts on maternal and newborn health in a variety of ways. On a large scale, the Millennium Development Goals 4 and 5 will not be met by 2015. Those countries which are least likely to meet the goals are the same countries with high levels of water insecurity and little likelihood of having secure and safe water resources any time soon (Requejo et al. 2014).

Water fetching jeopardizes the health and safety of young girls and women as a result of injuries from falls, animal attacks, and snake bites. Sexual assaults are common and a special risk for adolescent girls. The life and health of females are at risk in both situations. For women who are pregnant, these situations jeopardize continuation of the pregnancy and the safe delivery of healthy newborns.

Pre-pregnancy, women with chronic water-borne diseases suffer from vitamin deficiencies which compromise both becoming and sustaining a pregnancy. Water then plays a significant role in maternal mortality cause by complications of delivery (Benova et al. 2014) largely related to infections from unsafe delivery environments and postpartum care. Pregnancies resulting from the rape of young girls while water fetching results in obstructed labours, due to the small size of the girls' pelvises, and resultant caesarian sections performed in operating theatres lacking water and sanitary conditions. Similarly, newborns are compromised by lack of clean water for bathing and clean nursing environments both in health centres and at home.

Watt and Chamberlain argue that water insecurity is not a gender neutral issue. Women in their major role as mother are faced with special risk to their health in the pre-pregnancy, pregnancy, and postpartum period (Watt and Chamberlain 2011). Prevention is a key to healthy mothers and babies and, thus, should be a practice and policy concern of public health.

8.5.6 *Environmental Integrity*

Climate change is a relatively newly recognized threat to water security resulting in changes to patterns of infectious diseases, emerging and re-emerging pathogens, and consequently endangering public health. Harley et al. note that:

> Temperature, rainfall, humidity, and consequent physical and ecological characteristics of the environment set limits on the occurrence of a particular infectious disease. However, many other social, cultural, behavioral, technological, biological, and environmental factors act to determine where that infectious disease actually does occur (Harley et al. 2011).

On a global scale, the WHO released data regarding the estimated effects on human health as of the year 2000 (World Health Organization 2009). These data reveal that developing regions of the world have been disproportionately affected by climate change relative to developed regions. The impact of climate change, including water insecurity, has more severely compromised the already precarious population health in sub-Saharan Africa and Southeast Asia. WHO reports that "climate change is expected to increase storm surge-associated mortality in many regions of the world, in particular south Asia, high-income North America, Oceania, and east and west sub-Saharan Africa" (World Health Organization 2014b). In East Africa, malaria, a disease directly linked to excessive and stagnant water, has already migrated into the highland region, a locale previously malaria free (Shuman 2010). Although modern medicine is likely able to contain disease outbreaks and to curb mortality from these diseases, the impact on the health status of the population and the consequent social and economic damage should be of concern to public health authorities.

On a smaller scale, in Argentina, gated communities built on wetlands and floodplains in Greater Buenos Aires have modified fragile ecosystems aggravating flooding in poor surrounding neighbourhoods (Frayssinet 2014). Many of these wealthy private neighbourhoods have been built on floodplains and wetlands occupied by or adjacent to impoverished neighbourhoods. These ecosystems are vital to water drainage. This built environment induced flooding more frequently and more severely without increased rainfall that pose the same public health risks as flooding from other causes. Similarly it has been demonstrated that planned communities, such as Putrajaya, Malaysia can change patterns of infectious diseases such as dengue (Mulligan et al. 2012).

Urbanization has presented a particular challenge to providing water (Alirol et al. 2011). One aspect of the challenge has been the massive migration to urban centers in the least developed countries which have longstanding water security problems. This migration is largely unplanned in terms of infrastructure provision. Urban planning is required with a full understanding of the public health risks of failing to provide infrastructure to support water security (Tanner 2014).

Northern regions and mountainous regions in the south have become water insecure as a result of decreased total snowfall and prolonged periods of extreme cold that jeopardize access to water (Eichelberger 2014). The resultant lack of runoff to supply water reservoirs and damage to water transport systems from freezing temperatures threatens both access to and quality of water supplies in both urban and rural areas (Joshi and Tiwari 2014). Recent cold waves in Eastern Europe have led to frozen water pipes that resulted in no water availability for hospitals and health clinics (Byford 2014) and heat waves have produced life threatening dehydration of children, the sick, and the elderly and increased caregiver burden (Dominelli 2013). These instances make identifying, tracking, and intervening to prevent public health risks associated with water insecurity an important public health effort.

8.5.7 Disaster Management

Water insecurity results from damaging coastal events such storm surges and the increasingly intense tropical cyclones (Ramsay 2014) that result in significant death tolls and compromised agriculture, aquaculture, housing, and economic stability.

These disasters pose an immediate emergency public health challenge in terms of potable water scarcity and, where it previously existed, infrastructure damage and a longer term threat as a result of the physical and social disruption experienced by large numbers of people. Pollution from oil, gas, and chemicals from leeching and pipeline ruptures contaminates water in large land tracks putting populations of both urban and rural communities at risk (e.g. Hurricane Katrina in Louisiana) (Squires and Hartman 2013).

Breaching of levies which contain low lying water and the inadequacy of emergency resources such as population gathering points exposed people to disease and violence. (Gutmann et al. 2011) Both short and long-term mental health problem were caused by or exposed as a result of the hurricane especially among children and teens (Zwiebach et al. 2010). The latent inequities within the society in New Orleans based on race, class, and geographic location were identified as major determinants of outcomes – survival, resilience, and recovery post hurricane (Squires and Hartman 2013). In short, natural disasters such as hurricanes pose significant public health challenges not only to the physical environment but also to the physical and mental health of the population. When failure of water infrastructure compounds the damage of natural water disasters, planning of public health resources is critical to primary intervention, mitigation, and adaptation. Restoring health will be the goal of longer term strategies.

8.5.8 Population Growth

Currently at seven billion, the world's population is estimated to grow to 9.2 billion by 2050 (Bergstrom et al. 2013) placing additional strains on water resources, and threatening sufficient agricultural production to sustain that size of population (Cosgrove and Rijsberman 2014). Population growth is most accelerated in LMICs and characterized by significant urbanization that places even more pressure on already overtaxed water supplies. Crowded urban centers without clean water and with unsanitary environments are breeding grounds for infections and the spread of viruses.

In addition to absolute growth and internal migration, massive migrations related to natural and man-made disasters further threaten unstable water resources. Jordan, with advance water systems in an arid part of the world, has experienced water insecurity as migrants escaping wars in Syria and Mali (Bowles et al. 2014) increase the population of the country by 10% in a few months (UNHCR 2014a). Similar patterns have emerged in Kenya with refugees from Somalia (UNHCR 2014b). In both cases there has been serious stress on water resources resulting in threats to the health of residents not only in displaced persons camps, but also in surrounding communities.

8.5.9 *Population Safety During Armed Conflict*

Scheffran and Battaglini (Scheffran and Battaglini 2011) have identified water security as a critical element in climate change that plays an important role in creating security risks especially in parts of the world under water stress. They point out that cooperation is far more common than conflict when water is at risk and during natural disasters which transcend national boundaries. Although they acknowledge the possibility of conflict over water, they argue that this is really a by-product of already unstable or failing governments.

According to the World Water Assessment program "(T)the results from quantitative and qualitative analyses, based on science and economic principles, are often considered less relevant than political factors, emotion, religious beliefs and just gut feelings based on intuition" (WWAP 2012). As populations become overwhelmed and unable to adapt to changing resource provisions civil insecurity can emerge. The exact pathways to these ends are relatively unexplored and poorly understood.

A particularly precarious situation for both water and public health occurs when a community is at war. Destruction of community infrastructure including water infrastructure makes water scarce and unsafe. Recent examples include infrastructure destruction in Iraq and Bosnia (Micangeli et al. 2013; Olivera and Lewis 2004). There is a long history of regional conflicts over water sources and authors have gone as far as predicting the growing likelihood of "water wars" as the world supply of potable water decreases (Chellaney 2013). As mentioned earlier, major migrations of displaced persons further stresses the water supply (Zolnikov 2013) and the rebuilding of pre-existing water infrastructure in the post-war period will become both a priority and a significant burden on post-war economies (Micangeli et al. 2013; Weinthal et al. 2014).

8.5.10 *Health Information Dissemination*

Health information is essential for everyone in the population. Much of the information is about prevention and use of health services. Kreps notes that,

> Many of the people who are most at risk of poor health outcomes (…) are members of underserved populations, populations that are generally made up of individuals who are of low socioeconomic status, possess low levels of health literacy, are elderly, are members of marginalized ethnic and minority groups, or have limited formal education (Kreps 2005).

While the author was talking about the results of the Digital Divide Project in the U.S., he described the population in LMICs which have little access to health information through traditions media such as books, radio, television, or newspapers. Where digital media has taken hold, the nature of public health communication has changed the way in which public health information is communicated. Similarly, information about water insecurity could now be broadcast on cellphones to alert communities about impeding water related events. By coordinating with meteorologists and geologists, public health officials can predict geographic areas at high risk, alerting frontline staff of some types of upcoming events such as cyclones, snow storms, and tornados where time provides an opportunity to take preventive steps. Less time is available, at present, for events such as earthquakes and tsunamis where advance warnings are limited.

Even less warning is found for infrastructure failure. Flooding from broken water mains, contamination of water purification systems and sabotage of water systems happens under exigent circumstances that require immediate intervention to minimize damage and restore safe services. Hrudey, after examining several examples of water insecurity as a result of systems' failures, concluded that ongoing vigilance is required to ensure that water systems provide reliable, safe water even in HICs (Hrudey 2004).

Water systems need to be of concern to public health. Water treatment plants are the first line of public health defence in many communities (Schuster et al. 2005). In Canada as of October 31, 2016, there were 133 Drinking Water Advisories in effect in 90 First Nations communities across Canada, excluding British Columbia which no longer reports advisories through Health Canada (Health Canada 2016). Each of these advisories threatens, to varying degrees, the public health of the community in which the conditions exist.

8.6 Conclusions

Water insecurity and public health are inexorably linked through a number of pathways. Water insecurity undermines human security and threatens human health. It must be the job of water and public health practitioners, policy makers, and researchers to address these concerns beginning with the recognition that neither researchers, nor policy makers, nor practitioners can, in isolation of one from the other, fully describe, understand, or address the complex and entangled aspects of public health and water security. To do so will require transcending the traditional silos of water, sanitation, security, and emergency services, transportation, health care, human rights, and education. Some attempts have been made, with varying degrees of success, to bridge if not remove these divides with the goals of better managing complex public sector networks.

At a national level, the National Joint Sector Reviews for WaSH (Joint Sector Review: Getting the Basics Right 2014), of which Uganda (Hauge and Mackay 2004) and Liberia (Water, Sanitation & Hygiene (WASH) Sector Performance 2013 Report 2014) are leading examples, have brought key government stakeholders together around WaSH. In doing so, they expose the strengths of collaboration and cooperation and the challenges to traversing the barriers at both the local and country levels. These findings were confirmed in the reports of The Comprehensive Africa Agriculture Development Programme (Badiane et al. 2014) in which similar potential benefits and barriers to collaboration were identified in the agricultural sector. At an international level, Sanitation and Water for All combines high level dialogue and commitments with country level coordination and supporting research activities. In their recent document, A Global Framework for Action: guiding principles, they note, "The crosscutting nature of sanitation and water necessitates effective collaboration and joint working across relevant government ministries, donor and non-governmental agencies and other key institutions" (A Global Framework for Action – Governing Document 2013). The necessity and magnitude of the gaps to be spanned in order to achieve, integrated intersectoral, interdisciplinary, and cross-border research, policy, and practice within the water sector which includes public health is perhaps most evident when transboundary water issues are invoked. The Indus Waters Treaty focusing on the intersection of energy and irrigation (Indus Waters Treaty: Articles and Annexures 2007) and the Nile River Basin Commission focusing on planning, development, and management of water resources among the countries bordering on Lake Victoria (Mekonnen 2010) are two inter-sectoral international examples of bringing policy research and practice together. However, despite major health implications, health policy and practice is not even at the table.

In an era of hydro climatic uncertainty (Milly et al. 2008) education of both water and public health practitioners, researchers, and policy makers in the importance of integrated water health management that includes, but moves beyond, WaSH to encompass all productive uses of water is essential. One example of capacity development in this sphere is the UNU-INWEH on-line Water Learning Centre post-graduate diploma in integrated water-health management through Maseno University in Kenya (Integrated Water Resources Management 2015). Transdisciplinary staff working on the problems in practice, policy, and research are required as are institutionalised government co-ordination mechanisms. Both must become valued by funders and development overseers.

No one sector (or sub-sector in the case of water) has all the answers, or even all the questions. What is clear is that public health researchers, practitioners, and policy analysts must understand the problems of water security and come to the table as part of the solution. Equally, water specialists must recognize and respond to the public health issues in their sector. Motivation, resource efficiency, transparency, and accountability reduce the risk to third party investment thereby encouraging development opportunities in conjunction with the private sector. Modelling of cooperative, transdisciplinary, comprehensive approaches from both a top-down (e.g. UN Water) and bottom-up (e.g. NGO networks such as Uganda Water and Sanitation Network [UWASNET]). UN designated days that focus the world's attention on wicked problems, such as WaSH and public health, act as platforms to bring people to the table to brainstorm options and find solutions.

Lack of capacity, resource constraints, and political will are recurring roadblocks to cooperation and collaboration. If the Sustainable Development Goals are to live up to their potential, these systemic issues must be dealt with in innovative, cost neutral ways. To make real progress in bridging the water insecurity and public health divide, a proactive, transdisciplinary approach to prevention, mitigation, adaptation, and promotion must emerge. Ignoring or delaying this collaboration will come with significant costs to the health and well-being of communities world-wide and give new meaning to the line from Coleridge's 1789 poem The Rime of the Ancient Mariner – "Water, water everywhere and not a drop to drink" (Coleridge 2015).

Acknowledgement The author is grateful for the assistance of Dr. Corinne Schuster-Wallace, Senior Research Fellow, UNU-INWEH, in the preparation of this chapter. Her technical knowledge and passion for improving health, especially in rural and remote communities, enriches the global water security and public health dialogue.

References

A Global Framework for Action – Governing Document. (2013). Sanitation and water for all. New York: UNICEF.

Alexander, K. A., Carzolio, M., Goodin, D., & Vance, E. (2013). Climate change is likely to worsen the public health threat of diarrheal disease in Botswana. *International Journal of Environmental Research and Public Health, 10*(4), 1202–1230.

Alirol, E., Getaz, L., Stoll, B., Chappuis, F., & Louyan, L. (2011). Urbanisation and infectious diseases in a globalised world. *The Lancet Infectious Diseases, 11*, 131–141.

Allan, C., Xia, J., & Pahl-Wostl, C. (2013). Climate change and water security: Challenges for adaptive water management. *Current Opinion in Environmental Sustainability, 5*(6), 625–632.

Allegranzi, B., & Pittet, D. (2009). Role of hand hygiene in healthcare-associated infection prevention. *Journal of Hospital Infection, 73*(4), 305–315. doi: http://dx.doi.org/10.1016/j.jhin.2009.04.019

Anderton, D. L., & Leonard, S. H. (2004). Grammars of death an analysis of nineteenth-century literal causes of death from the age of miasmas to germ theory. *Social Science History, 28*(1), 111–143.

Ashraf, E., Shah, F., Luqman, M., Younis, M., Aziz, I., & Farooq, U. (2013). Use of untreated wastewater for vegetable farming: A threat to food safety. *International Journal of Agriculture and Applied Sciences, 5*(1), 27–33.

Azizullah, A., Khattak, M. N. K., Richter, P., & Häder, D. P. (2011). Water pollution in Pakistan and its impact on public health—A review. *Environment International, 37*(2), 479–497.

Badiane, O., Makombe, T., & Bahiigwa, G. (2014). Promoting agricultural trade to enhance resilience in Africa ReSAKSS annual trends and outlook report 2013. ReSAKSS Annual Trends and Outlook Report. Washington, DC.

Basu, M., & Shaw, R. (2013). Water scarcity and migration: An Indian perspective. *Community, Environment and Disaster Risk Management, 13*, 187–211.

Beaglehole, R., & Bonita, R. (1997). *Public health at the crossroads: Achievements and prospects*. Cambridge: Cambridge University Press.

Beaglehole, R., & Bonita, R. (2008). Global public health: A scorecard. *The Lancet, 372*(9654), 1988–1996.

Benova, L., Cumming, O., & Campbell, O. M. (2014). Systematic review and meta-analysis: Association between water and sanitation environment and maternal mortality. *Tropical Medicine & International Health, 19*, 368–387.

Bergstrom, R., Caddell, R., Chynoweth, M. W., Ellsworth, L. M., Henly-Shepard, S., Iwashita, D. K., et al. (2013). A review of solutions and challenges to addressing human population growth and global climate change. *International Journal of Climate Change: Impacts & Responses, 4*(3).

Black, R., Adger, W. N., Arnell, N. W., Dercon, S., Geddes, A., & Thomas, D. (2011). The effect of environmental change on human migration. *Global Environmental Change, 21*, S3–S11.
Black, R., Arnell, N. W., Adger, W. N., Thomas, D., & Geddes, A. (2013). Migration, immobility and displacement outcomes following extreme events. *Environmental Science & Policy, 27*, S32–S43.
Bouzid, M., Hooper, L., & Hunter, P. R. (2013). The effectiveness of public health interventions to reduce the health impact of climate change: A systematic review of systematic reviews. *PloS One, 8*(4), e62041.
Bowles, D.C., Reuveny, R., Butler, C.D.. (2014). Moving to a better life? Climate, migration and population health. In: *Climate change and global health* (pp. 135–143). Wallingford: CABI.
Bradley, D. J., & Bartram, J. K. (2013). Domestic water and sanitation as water security: Monitoring, concepts and strategy. *Philosophical Transactions of the Royal Society A: Mathematical, Physical and Engineering Sciences, 371*.
Braks, M. A., & De Roda Husman, A. M. (2013). Dimensions of effects of climate change on water-transmitted infectious diseases. *Air Water Borne Diseases, 2*, 2.
Byford, T. (2014). Protecting health in Europe from climate change. *International Journal of Environmental Studies* (ahead-of-print):1–2.
Chamberlain, G. R. L. P. (2007). *Troubled waters: Religion, ethics, and the global water crisis*. New York: Rowman & Littlefield Publishers.
Chellaney, B. (2013). *Water, peace, and war: Confronting the global water crisis*. Lanham: Rowman & Littlefield.
Cheng, J., Schuster-Wallace, C., Watt, S., Newbold, B., & Mente, A. (2012). An ecological quantification of the relationships between water, sanitation and infant, child, and maternal mortality. *Environmental Health, 11*(1), 4.
Coleridge, S. T. (2015). *The rime of the ancient mariner*. Mineola: Courier Dover Publications.
Cosgrove, W. J., & Rijsberman, F. R. (2014). *World water vision: Making water everybody's business*. Waltham: Routledge.
Cowell, C. (2001). The Hong Kong Fever of 1843: Collective trauma and the reconfiguring of colonial space. *Modern Asian Studies, 47*, 329–364.
Curtis, V. A., Danquah, L. O., & Aunger, R. V. (2009). Planned, motivated and habitual hygiene behaviour: An eleven country review. *Health Education Research, 24*(4), 655–673. doi:10.1093/her/cyp002.
Davis, M. J., Janke, R., & Magnuson, M. L. (2014). A framework for estimating the adverse health effects of contamination events in water distribution systems and its application. *Risk Analysis, 34*(3), 498–513.
Desai, A. (2013). Using a community-based participatory approach to build sustainable water and sanitation infrastructure to improve health status of northern Laotians. Paper presented at the 141st APHA Annual Meeting and Exposition (November 2–November 6, 2013).
Dingle, H. (1996). *Migration: The biology of life on the move*. New York: Oxford University Press.
Dominelli, L. (2013). Mind the gap: Built infrastructures, sustainable caring relations, and resilient communities in extreme weather events. *Australian Social Work, 66*(2), 204–217.
Ebi, K. L., Lindgren, E., Suk, J. E., & Semenza, J. C. (2013). Adaptation to the infectious disease impacts of climate change. *Climatic Change, 118*(2), 355–365.
Eichelberger, L. (2014). Spoiling and sustainability: Technology, water insecurity, and visibility in Arctic Alaska. *Medical Anthropology, 33*(6), 478–496.
Fee, E. (1993). Introduction. In G. Rosen (Ed.), *A history of medicine* (pp. i–lxviii). Baltimore/London: The John Hopkins Press.
Fenwick, A. (2012). The global burden of neglected tropical diseases. *Public Health, 126*(3), 233–236.
Field CBE. (2012). Managing the risks of extreme events and disasters to advance climate change adaptation: Special report of the intergovernmental panel on climate change: Special report of the intergovernmental panel on climate change.

Frayssinet, F. (2014). *Gated communities on the water aggravate flooding in Argentina*. New York: Inter Press Service News Agency.
Fung, I. C.-H., Fitter, D. L., Borse, R. H., Meltzer, M. I., & Tappero, J. W. (2013). Modeling the effect of water, sanitation, and hygiene and oral cholera vaccine implementation in Haiti. *The American Journal of Tropical Medicine and Hygiene, 89*(4), 633–640.
Gillispie, E. C., Sowers, T. D., Duckworth, O. W., & Polizzotto, M. L. (2015). Soil pollution due to irrigation with arsenic-contaminated groundwater: Current state of science. *Current Pollution Reports, 1*(1), 1–12.
Gundry, S., Conroy, R., & Wright, J. (2003). A systematic review of the health outcomes related to household water quality in developing countries. *Journal of Water and Health, 2*, 1–13.
Gutmann, A., Daniels, R. J., Kettl, D. F., & Kunreuther, H. (2011). *On risk and disaster: Lessons from Hurricane Katrina*. Philadelphia: University of Pennsylvania Press.
Habiba, U., Abedin, M. A., & Shaw, R. (2013). Defining water insecurity. *Community, Environment and Disaster Risk Management, 13*, 3–20.
Hall, P. (2014). *Cities of tomorrow: An intellectual history of urban planning and design since 1880* (4 ed.). West Sussex: Wiley-Blakcwell.
Hanjra, M.A., Ferede, T., Blackwell, J., Jackson, T., & Abbas, A. (2013). Global food security: facts, issues, interventions and public policy implications. Washington, D.C.
Harley, D., Swaminathan, A., & McMichaeL, A. J. (2011). Climate change and the geographical distribution of infectious diseases. In *Infectious diseases: A geographic guide* (pp. 414–423). Hoboken: Wiley Online Library.
Hauge, A. O., & Mackay, K. (2004). *Monitoring and evaluation for results: Lessons from Uganda*. Washington, DC: World Bank.
Headey, D. (2013). The global landscape of poverty, food insecurity, and malnutrition and implications for agricultural development strategies. Food Insecurity, and Malnutrition and Implications for Agricultural Development Strategies (November 2013).
Health Canada. http://www.hc-sc.gc.ca/fniah-spnia/promotion/public-publique/water-dwa-eau-aqep-eng.php. Accessed Dec 21, 2016.
Honeycutt, J., Hammam, O., Fu, C.-L., & Hsieh, M. H. (2014). Controversies and challenges in research on urogenital schistosomiasis-associated bladder cancer. *Trends in Parasitology, 30*(7), 324–332.
Hrudey, E. J. (2004). *Safe drinking water: Lessons from recent outbreaks in affluent nations*. London: IWA Publishing.
Indus Waters Treaty: Articles and Annexures. (2007). World Bank, Washington, DC.
Integrated Water Resources Management. (2015). United Nations Univresity Institute for Water, Environment and Health. http://wlc.unu.edu/courses/iwrm/. Accessed 13 Aug 2015.
IPCC. (2008). *Climate change and water*. Technical paper of the Intergovernmental Panel on Climate Change. Geneva.
IPCC. (2012). *Managing the risks of extreme events and disasters to advance climate change adaptation*. A special report of Working Groups I and II of the Intergovernmental Panel on Climate Change.
Jackson, S., Tan, P.-L., Mooney, C., Hoverman, S., & White, I. (2012). Principles and guidelines for good practice in indigenous engagement in water planning. *Journal of Hydrology, 474*, 57–65.
Joint Sector Review: Getting the basics right. (2014). Rural water supply network. Accessed 13 Aug 2015.
Joshi, B., & Tiwari, P.C.. (2014). Land-use changes and their impact on water resources in Himalaya. In: *Environmental deterioration and human health*. New York: Springer, pp. 389–399.
King, C. H. (2010). Parasites and poverty: The case of schistosomiasis. *Acta Tropica, 113*, 95–104.
Kouadio, I. K., Aljunid, S., Kamigaki, T., Hammad, K., & Oshitani, H. (2012). Infectious diseases following natural disasters: Prevention and control measures. *Expert Review of Infectious Diseases, 10*, 95–104.

Kreps, G. L. (2005). Disseminating relevant health information to underserved audiences: Implications of the Digital Divide Pilot Projects. *Journal of the Medical Library Association, 93*(4 Suppl), S68–S73.
Lobell, D. B. (2014). Climate change adaptation in crop production: Beware of illusions. *Global Food Security, 3*(2), 72–76.
Malm, A., Axelsson, G., Barregard, L., Ljungqvist, J., Forsberg, B., Bergstedt, O., & Pettersson, T. J. (2013). The association of drinking water treatment and distribution network disturbances with Health Call Centre contacts for gastrointestinal illness symptoms. *Water Research, 47*(13), 4474–4484.
McMichael, A. J. (2013). Globalization, climate change, and human health. *New England Journal of Medicine, 368*(14), 1335–1343.
Mekonnen, D. Z. (2010). The Nile basin cooperative framework agreement negotiations and the adoption of a 'water security' paradigm: Flight into obscurity or a Logical Cul-de-sac? *European Journal of International Law, 21*(2), 421–440. doi:10.1093/ejil/chq027.
Micangeli, A., Grego, S., Esposto, S.. (2013). Sustainable rehabilitation of water infrastructures in Southern Iraq after the Second Gulf War. In: *Handbook of sustainable engineering*. Dordrecht: Springer, pp 211–245.
Milly, P. C. D., Betancourt, J., Falkenmark, M., Hirsch, R. M., Kundzewicz, Z. M., Lettenmaier, D. P., & Stouffer, R. J. (2008). Stationarity is dead: Whither water management? *Science Communication, 319*, 573–574.
Moe C, Hurst C, Crawford R, Garkland J, Lipson D, Mills A, Stetzenbach L (2007) Waterborne transmission of infectious agents. In Manual of environmental microbiology. pp 222–248. Washington, DC: ASM Press.
Mulligan, K., Elliott, S. J., & Schuster-Wallace, C. J. (2012). Global public health policy transfer and dengue fever in Putrajaya, Malaysia: A critical discourse analysis. *Critical Public Health, 22*(4), 407–418. doi:10.1080/09581596.2012.659722.
Murray, C. J. L., & Lopez, A. D. (1997). Global mortality, disability, and the contribution of risk factors: Global Burden of Disease Study. *The Lancet, 349*, 1436–1442.
Naggar, Y. A., Naiem, E., Mona, M., Giesy, J. P., & Seif, A. (2014). Metals in agricultural soils and plants in Egypt. *Toxicological & Environmental Chemistry, 96*(5), 730–742.
Ngure, F. M., Reid, B. M., Humphrey, J. H., Mbuya, M. N., Pelto, G., & Stoltzfus, R. J. (2014). Water, sanitation, and hygiene (WASH), environmental enteropathy, nutrition, and early child development: Making the links. *Annals of the New York Academy of Sciences, 1308*(1), 118–128.
Nichols, G. L., Andersson, Y., Lindgren, E., Devaux, I., & Semenza, J. C. (2014). European monitoring systems and data for assessing environmental and climate impacts on human infectious diseases. *International Journal of Environmental Research and Public Health, 11*, 3894–3936.
Olivera, O., & Lewis, T. (2004). *Cochabamba!: Water war in Bolivia*. Brooklyn: South End Press.
Ramsay, H. (2014). Climate science: Shifting storms. *Nature, 509*(7500), 290–291.
Requejo, J.H., Bryce, J., Barros, A.J., Berman, P., Bhutta, Z., Chopra, M., Daelmans, B., de Francisco, A., Lawn, J., Maliqi, B. (2014). Countdown to 2015 and beyond: Fulfilling the health agenda for women and children. *The Lancet, 385*(9966), 466–476.
Rojias, P.M.. (2014). Haiti cannot wait 40 years. *Malta Times.*
Rosen, G. (1993). *A history of medicine*. Baltimore/London: The Johns Hopkins Press.
Sarkar, C., Webster, C., & Gallacher, J. (2014). *Healthy cities: Public health through urban planning*. Cheltenham: Edward Elgar Publishing.
Scheffran, J., & Battaglini, A. (2011). Climate and conflicts: The security risks of global warming. *Regional Environmental Change, 11*(1), 27–39. doi:10.1007/s10113-010-0175-8.
Schuster, C. J., Ellis, A. G., Robertson, W. J., Charron, D. F., Aramini, J. J., Marshall, B. J., & Medeiros, D. T. (2005). Infectious disease outbreaks related to drinking water in Canada, 1974–2001. *Canadian Journal of Public Health/Revue Canadienne de Sante'e Publique, 96*, 254–258.

Scott, C. A., Meza, F. J., Varady, R. G., Tiessen, H., McEvoy, J., Garfin, G. M., & Montaña, E. (2013). Water security and adaptive management in the arid Americas. *Annals of the Association of American Geographers, 103*(2), 280–289.
Shuman, E. K. (2010). Global climate change and infectious diseases. *New England Journal of Medicine, 362*(12), 1061–1063. doi:10.1056/NEJMp0912931.
Smith, K. (2013). *Environmental hazards: Assessing risk and reducing disaster*. London: Routledge.
Squires, G., & Hartman, C. (2013). *There is no such thing as a natural disaster: Race, class, and Hurricane Katrina*. New York: Routledge.
Sternberg, T. (2014). Transboundary hazard risk: The Gobi desert paradigm. *Natural Hazards, 72*(2), 533–548.
Tanner, M. (2014). *Urban health in developing countries: Progress and prospects*. London: Routledge.
Tuppad, P., Douglas-Mankin, K., Lee, T., Srinivasan, R., & Arnold, J. (2011). Soil and Water Assessment Tool(SWAT) hydrologic/water quality model: Extended capability and wider adoption. *Transactions of the ASABE, 54*(5), 1677–1684.
UNHCR. (2014a). 2014 UNHCR country operations profile – Jordan. http://www.unhcr.org/pages/49e486566.html.
UNHCR. (2014b). 2014 country operations profile – Kenya. http://www.unhcr.org/pages/49e483a16.html.
UNICEF, World Health Organization. (2012). *Progress on drinking water and sanitation: 2012 update*. Geneva: UNICEF, World Health Organization.
Vörösmarty, C. J., McIntyre, P. B., Gessner, M. O., Dudgeon, D., Prusevich, A., Green, P., & Davies, P. M. (2010). Global threats to human water security and river biodiversity. *Nature, 467*(7315), 555–561.
Water, Sanitation & Hygiene (WASH) Sector Performance 2013 Report. (2014). Ministry of Public Works, Monrovia.
Watt, S., & Chamberlain, J. (2011). Water, climate change, and maternal and newborn health. *Current Opinion in Environmental Sustainability, 3*(6), 491–496. doi:10.1016/j.cosust.2011.10.008.
Weinthal, E., Troell, J. J., & Nakayama, M. (2014). *Water and post-conflict peacebuilding*. New York: Routledge.
Winter, J. (2013). *London's teeming streets, 1830–1914*. New York: Routledge.
World Health Organization. (2006). *Meeting the MDG drinking water and sanitation target: The urban and rural challenge of the decade*. Geneva: World Health Organization.
World Health Organization. (2009). *Climate change and human health*. Geneva: World Health Organization.
World Health Organization. (2013a). *United Nations Children's Fund (UNICEF) Joint Monitoring Programme (JMP) for water supply and sanitation. Progress on sanitation and drinking water*. Geneva: World Health Organization.
World Health Organization. (2013b). *Sustaining the drive to overcome the global impact of neglected tropical diseases*. Second WHO report on neglected tropical diseases: summary.
World Health Organization. (2014a). *Investing in water and sanitation: Increasing access, reducing inequalities: UN-Water global analysis and assessment of sanitation and drinking-water GLAAS 2014*. Paris: Main Findings.
World Health Organization. (2014b). *Quantitative risk assessment of the effects of climate change on selected causes of death, 2030s and 2050s*. Geneva: World Health Organization.
WWAP (World Water Assessment Programme). (2012). *The United Nations world water development report 4: Managing water under uncertainty and risk*. Paris: UNESCO.
WWDR. (2012). *The United Nations world water development report 4: Managing water under uncertainty and risk*. Paris: UNESCO.
Yang, K., LeJeune, J. A., Alsdorf, D., Lu, B., Shum, C., & Ljang, S. (2012). Global distribution of outbreaks of water-associated infectious diseases. *PLoS Neglected Tropical Diseases, 6*, e1483.

Zolnikov, T. R. (2013). The Maladies of water and war: Addressing poor water quality in Iraq. *American Journal of Public Health, 103*(6), 980–987.
Zwiebach, L., Rhodes, J., & Roemer, L. (2010). Resource loss, resource gain, and mental health among survivors of Hurricane Katrina. *Journal of Traumatic Stress, 23*(6), 751–758.

Bibliography

Anstey, M. (2013). Climate change and health–What's the problem. *Globalization and Health, 9*(4), 1–5, doi:10.1186/1744-8603-9-4 http://www.biomedcentral.com/content/pdf/1744-8603-9-4.pdf.
Berrang-Ford, L., Dingle, K., Ford, J. D., Lee, C., Lwasa, S., Namanya, D. B., et al. (2012). Vulnerability of indigenous health to climate change: A case study of Uganda's Batwa Pygmies. *Social Science & Medicine 75*(6), 1067–1077. doi:10.1016/j.socscimed.2012.04.016.
Bircher, J., & Kuruvilla, S. (2014). Defining health by addressing individual, social, and environmental determinants: New opportunities for health care and public health. *Journal of Public Health Policy, 35*, 363–386.
Duffy, J. (1992). *The sanitarians: A history of American public health*. Chicago: University of Illinois Press.
Ford, J. D. (2012). Indigenous health and climate change. *American Journal of Public Health, 102*, 1260–1266.
Furgal, C., & Seguin, J. (2006). Climate change, health, and vulnerability in Canadian northern Aboriginal communities. *Environmental Health Perspectives*, suppl., 1964–1970.
Guest, C., Ricciardi, W., Kawachi, I., & Lang, I. (2013). *Oxford handbook of public health practice*. Oxford: Oxford University Press.
Huang, C., Vaneckova, P., Wang, X., FitzGerald, G., Guo, Y., & Tong, S. (2011). Constraints and barriers to public health adaptation to climate change: A review of the literature. *American Journal of Preventive Medicine, 40*(2), 183–190, doi:10.1016/j.amepre.2010.10.025.
Kjellstrom, T., & Weaver, H. J. (2009). Climate change and health: Impacts, vulnerability, adaptation and mitigation. *New South Wales Public Health Bulletin, 20*(2), 5–9.
Marmot, M., Friel, S., Bell, R., Houweling, T., & Taylor, S. (2008). Closing the gap in a generation: Health equity through action on the social determinants of health. *The Lancet, 372*(9650), 1661–1669, doi:10.1016/s0140-6736(08)61690-6.
Mboera, L. E., Mfinanga, S. G., Karimuribo, E. D., Rumisha, S. F., & Sindato, C. (2014). The changing landscape of public health in sub-Saharan Africa: Control and prevention of communicable diseases needs rethinking. *The Onderstepoort Journal of Veterinary Research, 81*(2), 1–6.
Pawar, M. (2013). Water insecurity: A case for social policy action by social workers. *Australian Social Work, 66*(2), 248–260.
Rosen, G. (1993). A history of medicine. Baltimore/London: The Johns Hopkins Press.
Séguin, J., Berry, P., Bouchet, V., Clarke, K. L., Furgal, C., Lamy, S., … & MacIver, D. (2008). Human health in a changing climate: A Canadian assessment of vulnerabilities and adaptive capacity. *Human Health in a Changing Climate, 1*.
Turnoch, B. J. (2009). *Public health: What it is and how it works*. Sudbury: Jones & Bartlett Publishers.
Tulchinsky, T. H., & Varavikova, E. A. (2000). History of public health. In *The New public health: An introduction for the 21st century* (pp. 1–54) Academic Press.
UNICEF, & World Health Organization. (2011). Drinking Water Equity, Safety and Sustainability: Thematic report on drinking water 2011.

United Nations (UN). (2008). *61st world health assembly climate change and health*. Geneva: United Nations.
UN-WATER. (2013). Analytical Brief on Water Security and the Global Water Agenda.
World Health Organization. (2010). *Progress on sanitation and drinking water 2010 update*. Geneva.
World Water Assessment Programme (WWAP). (2014). *The united nations world water development report 2014: Water and energy*. Paris: UNESCO.
Zeitoun, M. (2011). The global web of national water security. *Global Policy, 2*(3), 286–296.

Chapter 9
Going to the Well: Water as a Community Builder

Jennifer Fresque-Baxter and Erin Kelly

Abstract Communities – both fixed and fluid – play a critical role in water stewardship and protection of freshwater. In this chapter, we examine the development and implementation of the Northwest Territories (NWT) Water Strategy as a mechanism for promoting collaborative action towards addressing water security challenges. Key examples of water-related initiatives and experiences in addressing water security at multiple levels are explored. A series of insights for global practitioners, based on our experiences in implementing the Water Strategy, are offered, and framed within respective successes and challenges. The chapter concludes with a discussion on the potential utility of thinking about framing 'community' from the lens of 'communities of practice', and future research directions in this regard.

Keywords Water stewardship • Water security • Northwest Territories • NWT Water Strategy • Collaboration • Communities of practice

9.1 Introduction

Water often crosses multiple jurisdictional boundaries, sectors, stakeholder interests and political and agency mandates (Blomquist and Schlager 2005; Dunn et al. 2012). Advancing practices that support equitable access to safe, clean, abundant waters for people and ecosystems requires inputs from multiple actors and stakeholders with varied interests at multiple levels and scales (de Loë et al. 2007; UN-Water 2013) and should be positioned "within a model that promotes good

J. Fresque-Baxter (✉) • E. Kelly
Department of Environment and Natural Resources, Government of the Northwest Territories, Yellowknife, NT, Canada
e-mail: Jennifer_fresque-baxter@gov.nt.ca

D. Devlaeminck et al. (eds.), *The Human Face of Water Security*, Water Security in a New World, DOI 10.1007/978-3-319-50161-1_9

governance" (Norman et al. 2010, p. iii). 'Good' water governance should be transparent, accountable, integrative and participatory, among other principles (de Loë et al. 2007; Norman et al. 2010).

Cook and Bakker (2012, p. 100) and Dunn et al. (2012) advocate for a 'broad, integrative' approach to defining water security. Cook and Bakker (2012) and Dunn et al. (2012) note that an integrative framing can be more comprehensive, in that a broader suite of factors that shape water security and its management may be reflected.[1] Thus, in this chapter, we approach water security broadly, as multi-dimensional, and in an integrated fashion, after approaches advocated by key national and international scholars and practitioners (e.g., Cook and Bakker 2012; Dunn et al. 2012; de Loë et al. 2007; Global Water Partnership 2012; Grey and Sadoff 2007).

Water is fundamental to residents of Northern Canada for cultural, subsistence, economic and recreational purposes (Government of the Northwest Territories [GNWT]/Indian and Northern Affairs Canada [INAC] 2010). Concerns about changes in water quality and quantity, and related impacts on human and aquatic ecosystem health, led to the release of *Northern Voices, Northern Waters: NWT Water Stewardship Strategy* ('Water Strategy') in 2010. The concerns related to water security raised by Northerners during the development of the Water Strategy ranged from impacts at the local level to regional, territorial and transboundary levels. The Strategy was a collaborative endeavor between Aboriginal, municipal, territorial and federal governments, communities, industry, environmental non-government organizations (ENGOs) and academics. The common vision for Northern waters that emerged out of the development of the Strategy is that "*waters of the NWT will remain clean, abundant and productive for all time*" (GNWT/INAC 2010, p. 10). As outlined in the Water Strategy, waters must remain this way to ensure: (1) human health and access to safe drinking water at all times; and, (2) that aquatic ecosystems are healthy, among other goals. From our perspective, the vision and its related goals represent the epitome of water security.

It is our position that the experiences from the development and implementation of the Water Strategy – and how related activities can function to support water security in the NWT – can offer broad insights for practice and governance at regional (i.e., Basin/watershed-wide levels), national and international levels.

9.2 The Mackenzie River Basin and Northwest Territories

Much of NWT is part of the Mackenzie River Basin (MRB), Canada's largest watershed at 1.8 million km^2 – comprising roughly 20% of Canada's land mass – and the tenth largest river basin globally (Rosenberg International Forum on Water Policy 2013; MRBB 2003). The Basin encompasses parts of the provinces of Saskatchewan,

[1] Though both the aforementioned papers note that 'narrow' approaches can complement broad framings and may have roles to play around operationalizing water security at the program level. For further discussion, see Dunn et al. (2012) and Cook and Bakker (2012).

Alberta and British Columbia, Yukon Territory and NWT (Rosenberg International Forum on Water Policy 2013; MRBB 2003), spanning geopolitical boundaries. The Mackenzie River contributes roughly 60% of all freshwater discharge from Canada to the Arctic Ocean, playing a major role in regulation of climate system variability and ocean circulation (Mackenzie Global Energy and Water Cycle Experiment Study 2004; MRBB 2003; Palmer et al. 2012; Rosenberg International Forum on Water Policy 2013). Thus, protection of water and aquatic ecosystem health and function of the MRB is of paramount importance not only for the NWT and Canada, but globally (Rosenberg International Forum on Water Policy 2013). The MRB is home to around 400,000 people, most of whom live within its southern portions (i.e., below the 60th parallel) (Palmer et al. 2012). The waters of the MRB are culturally significant to Aboriginal peoples residing within it, support livelihoods, traditional and subsistence activities and the interconnected plants, animals, fish, land and air that people rely on (Rosenberg International Forum on Water Policy 2013). NWT also shares waters with the territory of Nunavut.

In Canada's North, discussions about water are not separated from other aspects of the environment, including people, animals, air, land and climate. Furthermore, water security is inextricably linked with related notions of food and livelihood security and must also be situated, and understood, within the context of climate change, energy and resource development (see for example, Allan 2013; Froggatt 2013; Ludi 2009; World Economic Forum 2011). Thus, security and stewardship of water are embedded within broader ecological, social, political, economic and cultural contexts. However, water is often a natural entry point for framing and discussing environmental change, management and stewardship. A common adage in the NWT and MRB, largely stemming from and reflective of Aboriginal people's relationships with water, is that 'water is life' (Aboriginal Affairs and Northern Development Canada 2010; NWT Water Stewardship n.d.). Though focus in this chapter is on water security specifically, and management and practices that support security, the consideration of its embedded nature (i.e., linkages to food, climate, energy, etc.) is always central in our minds.

9.3 Northern Voices, Northern Waters: Northwest Territories Water Stewardship Strategy Development and Implementation

Residents of the NWT have expressed concerns about water and identified potential threats to water security within the NWT and MRB for decades, particularly those related to impacts from growth, development and climate change. Prior to the development of the Water Strategy, a series of Aboriginal-led conferences occurred to share experiences and knowledge regarding water-related concerns. Examples of these events include: Water Wise Conference (2007), Keepers of the Water Gatherings (2006, 2007, 2008), Sahtu Water Gathering (2008) and the National Summit on the Environment and Water hosted by the Dene Nation in 2008 (GWNT/INAC 2010).

In March 2007, Members of the 15th Legislative Assembly of the Government of the Northwest Territories (GNWT) passed a motion stating "water is essential to life, and constitutes a fundamental human right" (as cited in GWNT/INAC 2010, p. 44). Recognition of this right is consistent with many international rights declarations and plans of action (e.g., general comment No. 15 of the UN Committee on Economic, Social and Cultural Rights; Agenda 21, etc.) (Office of the High Commissioner on Human Rights et al. 2010).

Also in 2007, the GNWT committed to developing a NWT water resource management strategy, in response to concerns raised at the aforementioned gatherings and the recognition of water as a human right. *Northern Voices, Northern Waters: NWT Water Stewardship Strategy*, released in 2010, was the result of collaboration and engagement with Aboriginal leadership, communities, governments at multiple levels, regulatory boards, ENGOs, academics and industry (actors and stakeholders commonly referred to in the NWT as 'water partners', which will henceforth be used in this chapter) through a series of workshops, meetings and presentations. To ensure action and accountability, *NWT Water Stewardship: A Plan for Action 2011–2015* was released in 2011 and outlined key actions/deliverables, timelines and lead partners. Critical to the success of the Water Strategy was the creation of an Aboriginal Steering Committee (ASC) to guide its development and implementation. The ASC ensures Aboriginal culture, rights, values and knowledge, as well as community and regional needs and priorities, are considered and respected throughout Water Strategy implementation.[2]

The Water Strategy reflects the fundamental relationship that exists between NWT residents and water. The vision, "that the waters of the NWT remain clean, abundant and productive for all time" (GNWT/INAC 2010, p. 10) and goals of the Water Strategy set out targets for NWT waters, which include ensuring aquatic ecosystems remain healthy and diverse, and humans, as part of the ecosystem, can choose their way of life and have access to clean and abundant water for drinking, food, travel, economic growth, culture and spirituality now and for future generations. The Water Strategy supports the "appropriate use and consideration of all types of knowledge including traditional,[3] local and western scientific" for aquatic ecosystem-related decision-making (GNWT/INAC 2010, p. 4). Ultimately, the Water Strategy "encourages responsible economic development within a sound environmental context" (GNWT/INAC 2010, p. 3), to further support secure waters for people and ecosystems, now and for future generations.

---

[2] Current ASC membership includes representatives of the seven regional Aboriginal governments (including one with observer status), individual Aboriginal governments who are not part of a regional Aboriginal government, and the territorial government.

[3] In the NWT, 'traditional knowledge' has been defined as "knowledge and values, which have been acquired through experience, observation, from the land or from spiritual teachings, and handed down from one generation to another" (GNWT Traditional Knowledge Policy 2005, p. 2). In the academic literature and in other places around the world different terms may be used, such as 'traditional ecological knowledge' and 'Indigenous knowledge' (see for example, Berkes 2008; Berkeset al. 2000). In this chapter, we use 'traditional knowledge', as that is the term commonly used in the NWT.

One of the most important aspects of the Water Strategy to Aboriginal peoples within the NWT is that its development and implementation do "not affect or infringe upon existing or asserted Aboriginal rights, treaty rights or land resource and self-government agreements. In the case of any inconsistency between the Water Strategy and existing or future treaties or land, resource and self-government agreements, the provisions of the treaties and agreements shall prevail (GNWT/INAC 2010, p. 4)."

The Water Strategy's guiding principles are respect, sustainability, responsibility, knowledge and accountability. Approaches to meet Water Strategy goals include stewardship, ecosystem-based management, water and watershed values, and information to understanding. These guiding principles and approaches, coupled with 'Keys to Success' and identified actions, define the agreed upon path toward improved and strengthened water stewardship and water security in the NWT. Keys to Success, which are "what we need to do" to achieve the vision and the goals of the Water Strategy, exist under four categories: (1) 'Work Together'; (2) 'Know and Plan'; (3) 'Use Responsibly'; and, (4) 'Check Our Progress'. Specific action items for each Key to Success are detailed within the Action Plan with associated leads and timelines documented to ensure accountability. Please see Table 9.1 for an overview of each Key to Success category, with select examples. The majority of the examples listed are described in greater detail throughout the chapter, either in the main text or in a series of text boxes.[4] In some of the text boxes, we have also included input provided by water partners involved at various levels of Water Strategy implementation to provide additional depth and experience, particularly related to achieving water security in the NWT.

The Water Strategy Action Plan is a living document. The path forward is continually updated based on the learning and experience of all partners together. Annually, water partners gather to discuss progress and implementation of Strategy activities. These meetings provide an opportunity to continually assess and (re)affirm priorities for coming years. Occasional reports cards also highlight areas where Strategy work has progressed and where room for improvement exists.

Every five years, an independent, third-party evaluation is conducted to objectively assess progress towards achievement of Water Strategy goals and actions. This important check and balance helps ensure activities continue to meet Northerner's needs and ensures water partners are accountable for the implementation of the Strategy. The evaluation determines where successes can be built upon and where there are key areas for improvement. The 2015 independent evaluation of the Strategy found that excellent progress had been made on negotiating transboundary agreements with neighbouring jurisdictions; increasing the number of communities involved in community-based monitoring; and, achieving objectives related to building meaningful partnerships to support water stewardship. The evaluation also found that there were areas of importance to water partners that require improvement. For example, while progress has been made on community-based monitoring, expansion to additional interested communities and finding ways to retain trained community water quality monitors over the long term is needed. The need to more formally

[4] For more details on these, and other, implementation activities, please visit nwtwaterstewardship.ca

Table 9.1 Catergories of Keys to Success

Category	Work together	Know and plan	Use responsibly	Check our progress
Brief description	Building a cooperative space for water partners to collaborate	Collectively understanding ecosystem health through multiple ways of knowing	Ensuring up-to-date, relevant policy and guidance for informing water use decisions	Evaluating the progress of implementation through routine checks and formal audits
Core desired outcomes	-Improved collaboration -Improved communication -Increased awareness and engagement -Access to and use of best available information for informed decisions at multiple scales	-Collectively developed comprehensive monitoring and research programs -Opportunities for active and meaningful community involvement -Holistic understanding of ecosystem processes and health -Timely and accessible reporting to inform decision-making	-Consistency and transparency in decision-making -Routine evaluation of legislation and regulations to maintain relevancy and appropriateness -Ensuring capacity to promote compliance	-Transparency and accountability of Strategy implementation -Adaptability to changing priorities and circumstances (e.g., climate change) -(Re)affirming partner commitment -Relevancy to northern partner needs
Examples of actions/ activities	-Bilateral Water Management Agreements -Mackenzie DataStream data management system	-NWT-wide Community-based Water Quality Monitoring Program -Dehcho First Nations AAROM program -Slave River and Delta Partnership	-Completion of *The Guidelines for the Closure and Reclamation of Advanced Mineral Exploration and Mine Sites in the NWT* -Completion of policy document on Water and Effluent Quality Management -ENR-GNWT review of Water License submissions from proponents in the spirit of the Water Strategy	-Annual water partner workshops -Progress report cards -Independent evaluation (2015)

define who constitutes a water partner, identify their roles and responsibilities in Water Strategy implementation and improve collective ownership over the Water Strategy were also identified as areas requiring additional effort.

These successes and challenges are currently being addressed and point to important new directions for the Strategy's future through development of a new five-year Action Plan, collaboratively developed during 2015. Lessons learned from Water Strategy implementation between 2011 and 2015 and results of ongoing evaluation informed the development of the next Action Plan as did guidance from the ASC and substantial engagement with all water partners.

9.4 Insights from Experience: Advancing Water Security Through the NWT Water Strategy

Water binds people in the NWT together across environmental, social, cultural and political contexts. Water is life and sustaining it requires sharing and learning among people with a vested interest in protecting it now and into the future. The Water Strategy has built on the concerns and priorities of NWT residents, to achieve its goals – which strongly reflect maintaining water security for people and the environment. The Water Strategy brings together people who share resources, tools and practices to advance water stewardship towards the vision of safe, sustainable waters in a continually evolving and dynamic approach to water stewardship (see also discussion on 'Communities of Practice' below).

Below are key insights from our experiences in implementing the Water Strategy we feel may be pertinent for other practitioners who are dealing with similar challenges and contexts. The key themes identified below also reflect traits that can foster resilience in the governance of aquatic systems (see for example, Plummer et al. 2014). It is important to note, however, that the Strategy is very much grounded in a Northern Canadian experience, though in many respects it is consistent with international practices and declarations. We feel our experiences – including our challenges and successes – may resonate with global practitioners attempting to engage in collaborative watershed management, to link multiple knowledge perspectives (e.g., indigenous knowledge and western science) and to address cumulative impacts on aquatic ecosystems, including those from climate change and resource development.

9.4.1 Common Vision, Common Goals

Central to the success of the Water Strategy was the development of a common vision and goals to guide water stewardship and management. As noted by Mitchell (2006), in integrated water resources management (IWRM), having a vision provides an end point for people to work towards and can help to determine the people and elements of the system that need to be brought together to reach that outcome. As Dunn et al. (2012) and Cook and Bakker (2012) note, water security can potentially be such a vision. The Water Strategy vision and goals reflect the needs and priorities of Northerners, as they were collaboratively developed. In the case of the Water Strategy, the vision reflects the very epitome of water security as the outcome of collective action and ecosystem-based management.

It is also important to note that water partners have their own interests, needs and mandates, and there are times when some water partners may not clearly identify with the Water Strategy vision and goals. This has represented a challenge in Water Strategy implementation. For example, the first external evaluation noted the need to increase feelings of ownership of all water partners in Water Strategy implementation. Some partners also identified that certain implementation areas or activities have been prioritized over others, causing uncertainty related to how best to engage in work on actions not deemed as high priority under the Strategy but of importance to particular water partners.

To address such challenges, the Water Strategy is framed as a living document and reflects the need for adaptability to redefine priorities of water partners as they may change or as more information is gathered. What we have learned through implementation, is the necessity of continually and collectively revisiting the goals and priorities – and the activities needed to achieve those – to ensure they remain relevant and meaningful to all NWT water partners. Progress has been made through participation in collaborative meetings and improved communications during the past five years of implementation. For example, as previously described, annual water partner meetings are held where results and updates from ongoing projects are presented and activities for the coming year are prioritized. External evaluations also provide an important check and balance to ensure that activities reflect current needs and priorities of all water partners.

9.4.2 Knowledge and Learning

Knowledge is one of the guiding principles of the Water Strategy and "decisions are based on accurate and up-to-date traditional, local and western scientific knowledge" and "as knowledge evolves, stewardship decisions evolve accordingly" (GNWT/INAC 2010, p. 11). For all water partners to make good decisions about water stewardship and management – and work towards water security as an outcome – they must have access to the best available information (from multiple sources of knowledge) to support the choices they make.

From our experience, one of the critical elements of informed decision-making is ensuring people have access to the information they need in ways that are mean-

Box 9.1 Transboundary Water Management Agreement Negotiations

The NWT is the ultimate downstream jurisdiction in the Mackenzie River basin. As such, impacts to upstream waters can affect downstream users in the NWT. In 1997, the *Mackenzie River Basin Transboundary Waters Master Agreement* ('Master Agreement') was signed by the governments of Alberta, British Columbia, Saskatchewan, Yukon, the NWT and Canada. It committed the Parties to work together to cooperatively manage the waters of the MRB "in a manner consistent with the maintenance of the ecological integrity of the aquatic ecosystem" (Master Agreement 1997, p. 3). A series of bilateral water management agreements (BWMA) were to be negotiated under the Master Agreement. The Yukon-NWT BWMA was signed in 2002, the Alberta-NWT BWMA in March 2015 and the British Columbia-NWT BWMA in October 2015. The NWT is currently involved in negotiations toward a BWMA with Saskatchewan. Discussions related to updating the existing BWMA with Yukon have begun. Currently, MRB jurisdictions are in general agreement on a 'base ecosystem agreement' that includes provisions related to: risk informed management, surface water quality and quantity, groundwater, biology, monitoring, research, information sharing, notification and consultation, emergency response administration, dispute resolution and process provisions. Comprehensive appendices accompany BWMAs and include detailed technical information, a path forward for using traditional and local knowledge during implementation, learning plan contents and cost sharing information.

For the NWT, the discussions on the development of the Water Strategy informed the NWT's BWMA negotiations mandate. Thus, the BWMAs are consistent with input NWT residents provided in terms of water stewardship. Traditional and local knowledge of transboundary waters was also gathered through engagement and consultation with NWT Aboriginal groups to help inform NWT's negotiation interests. The ASC was an integral part of negotiating these agreements and will continue to be actively and meaningfully involved throughout implementation.

ingful, accessible and culturally- and locally-appropriate. This can mean tailoring communications to meet different user needs, experiences and values, languages[5] and ensuring the information is relevant to the receiver (contextualized). If the information is not reaching partners, is not relevant to their interests and needs or is not in a format that can be used by those partners (for many reasons), it may not be used effectively to support decision-making. Finding appropriate mechanisms for reciprocal sharing of information is something the Department of Environment and Natural Resources (ENR-GNWT) and many water partners have worked to improve and is an important aspect of building capacity for water stewardship.

[5] The NWT recognizes 11 official territorial languages, nine of which are Aboriginal languages.

Of paramount importance for the NWT is bringing multiple knowledges to bear in water governance. Indeed, as Armitage et al. (2015) note, "contemporary water governance must draw on knowledge in its many different forms (scientific, local, indigenous, bureaucratic)" (p. 361). The Water Strategy is predicated on the use of best available information for decision-making, which is grounded in the need to draw upon multiple knowledge sources. As noted in the Water Strategy, "Traditional, local and western scientific knowledge all contribute to understanding these aquatic ecosystems and their stressors, such as climate change and human development" (GNWT/INAC 2010, p. 19). In the NWT, Aboriginal peoples have knowledge of and relationships to the water and the land that have existed since time immemorial. Such knowledge is a "a valid and essential source of information about the natural environment and its resources, the use of natural resources, and the relationship of people to the land and to each other" (GNWT Traditional Knowledge Policy 2005, p. 1). Multiple knowledges were fundamental to the development of the Alberta-NWT Bilateral Water Management Agreement (see Box 9.1). The work of the Slave River and Delta Partnership, including through its SWEEP project (see Box 9.2), is centered on the blending of multiple knowledge systems to understand the health of the Slave River and Delta.

Bringing together multiple knowledge systems has unique challenges, which include the tendency to privilege western scientific information in resource management and research (Berkes 2008; Adams et al. 2014). In our experience, working to try and bridge knowledge systems requires collaborative, reciprocal and respectful relationships between partners (see also Adams et al. 2014, for more discussion with respect to partnerships between academic and Indigenous partners in ecological research). It can be affected by a host of factors including capacity of water partners, diverse and competing interests, misunderstandings between perspectives and geographic distance to name a few. In our experience, commitment to listening and understanding, building strong partnerships, collectively defining goals and values at the outset of projects, establishing trust and undertaking sustained, meaningful interactions as often as possible, all serve to create the space for dialogue that fosters knowledge exchange.

Within the Water Strategy, traditional knowledge is recognized as equal and parallel with western science. This counters potential privilege of one form and source of knowledge over another – though we strongly recognize the need to continually work towards improvements in this regard throughout implementation. Water partners continue to work collectively to improve the processes by which traditional and local knowledge inform decision-making at all levels. Partners are also looking for opportunities to improve blending of multiple knowledges and to collectively create knowledge (that draws on multiple knowledges) to support water stewardship and management (the concept of 'knowledge co-production', *sensu* Tengö et al. 2014; Armitage et al. 2011, 2015; Pohl et al. 2010; see also Berkes 2008). As one means of addressing this priority, a number of Aboriginal governments from throughout the MRB, the GNWT, Mackenzie River Basin Board (MRBB) and a number of national and international academic institutions, recently partnered with the University of Alberta on a six-year project called 'Tracking Change', focused on the role of traditional and local knowledge in watershed governance in the Mackenzie, Mekong and Amazon River Basins.[6] ENR-GNWT continually works with water partners to identify the appropriate mechanisms for gathering information (particularly local and traditional knowledge) in ways that honour and respect the knowledge holders.

Adaptive management is positioned strongly in the Water Strategy and is part of the very nature of this living document. It is particularly important for addressing issues of water security, as new challenges (e.g., climate change, development, socio-economic impacts) may result in changes to how (in)secure NWT water is; therefore, our management approaches and structures must be flexible enough to account for, and adapt to, changing circumstances in order to remain effective.

Box 9.2 Slave River and Delta Partnership

During the fall of 2010, residents in Fort Smith began to catch pickerel that had bumps on them from the Slave River, which is located downstream of the Athabasca oil sands, during one of the lowest water years on record. This was very concerning for community members who continue to rely on harvesting fish. It was clear more research and monitoring needed to be done and the communities along the Slave River and Delta needed and wanted to drive the design and implementation of the work. It was from this that the Slave River and Delta Partnership (SRDP) was formed. The SRDP consists of Aboriginal, municipal, territorial and federal governments and northern educational institutions.

SRDP research and monitoring is guided by three overarching questions, which were identified by the community partners: *Can we drink the water? Can we eat the fish? Is the ecosystem healthy?* Upon further discussion, many community members questioned whether their children and grandchildren would be able to drink the water, eat the fish and choose a subsistence lifestyle. All of these questions are related to water and food security. From the very beginning, the group strongly identified the importance of answering questions by drawing on both traditional knowledge and western science together to ensure a holistic perspective on the health of the Slave River and Delta. Academics from various Canadian universities and some ENGOs have worked with and supported the SRDP to undertake a variety of projects to address their questions.

Most recently, an interdisciplinary team of University of Saskatchewan researchers has been working closely with the SRDP to conduct scientific and traditional knowledge research and monitoring toward developing an

(continued)

[6] The project, led by principal investigator Dr. Brenda Parlee (University of Alberta), is funded by the Social Sciences and Humanities Research Council, one of the Canadian Federal Tri-Council research agencies.

Box 9.2 (continued)
Aboriginal-led community-based cumulative effects monitoring program for the Slave River and Delta. This project, called the Slave Watershed Environmental Effects Program (SWEEP), has expanded on aquatic ecosystem health indicator work previously completed by the SRDP and is funded by the Canadian Water Network. Research and monitoring undertaken by the SRDP, particularly through SWEEP, has resulted in increased information on the health of the Slave River, training opportunities for community partners to build capacity to monitor a variety of parameters and improved reciprocal learning and knowledge sharing between partners.

9.4.3 Collaboration and Participation

Partnerships are central to activities under the Water Strategy for many reasons, as evidenced throughout this chapter. In order to address large, complex and multi-scale problems – such as promoting water security – there is a need to pool resources (including financial and human capital) and work collectively to maximize impact and capacity. There are many people with a stake in NWT water stewardship and all should be involved in developing the solutions that move the NWT towards water security. Partnerships have been critical in developing community-based monitoring programs in the NWT (see Boxes 9.3 and 9.4 for descriptions of two programs). When you bring together diverse partners, with different knowledges, values and experiences, you get a variety of perspectives that can serve to strengthen approaches or identify innovative solutions. Furthermore, given the common interests, and limited capacity, it is important to avoid duplicating efforts or 'reinventing the wheel'.

Box 9.3 Aquatic Monitoring in the Dehcho Territory, by George Low, Dehcho Aboriginal Aquatic Resources and Oceans Management (AAROM)
The Dehcho Aboriginal Aquatic Resources and Oceans Management (AAROM) program is a community-based capacity building and aquatic resource management program partnering with nine First Nations and two Métis organizations in the Dehcho region of the Northwest Territories. The program, funded by Fisheries and Oceans Canada, provides a bridge to the more formal resource management responsibilities First Nations will take on once they have settled land claims or treaty entitlements.
The Dehcho AAROM mission is for "more Aboriginal control of fish and water resources under the Dehcho First Nation's 'one house' system of

(continued)

Box 9.3 (continued)
governance – involving local resource users and enhancing employment and educational opportunities for Dene Youth." Our mandate is to "protect and preserve the rights of all Aboriginal people to healthy waters, fish stocks and aquatic environments in the Dehcho." The Dene and Métis people of the Dehcho have always expressed a deep rooted respect for the land and waters. The Dehcho Declaration includes the statement "we were put here by the Creator as keepers of the waters and lands." Most Dene and Métis have now moved off the land and traded the traditional lifestyle for the wage economy in the settlements. Hunting, trapping and fishing still provide healthy traditional food and these activities are central to the cultural and spiritual well-being of the Aboriginal people of the Dehcho.

The Dehcho territory is blessed with an abundance of fresh water as part of the Mackenzie River Basin, however, there are concerns about water quality and the presence of contaminants from upstream southern reaches of the watershed including pollution from tar sand projects, pulp and paper mills, coal mines and other potential sources of industrial waste. There is also a concern with water quantity. The Mackenzie River was recently at record low levels and many of the tributaries are running dry. Travelling on the River to hunt and fish has become more difficult and it has become impossible to travel up tributaries such as the Trout, Redknife and Rabbitskin rivers. The W.A.C. Bennett Dam has changed seasonal water-flow and water-level regimes since it was built. There is a concern that additional dams planned on the Peace River will cause greater problems. First Nations in the Dehcho are also worried about the effects of greater water use on the Athabasca River as well as apparent climate change effects.

To begin addressing these concerns, the Dehcho AAROM program has been building a community-based water monitoring program in the Dehcho region. Recently, this program has been greatly strengthened through partnering with ENR-GNWT as a partner in the NWT Water Strategy. Water quality measuring instruments as well as devices for measuring petrochemical contaminants and metals are deployed and maintained during the summer months near several communities. Dehcho AAROM and its community partners, through a funding agreement with ENR-GNWT are responsible for providing the collected data for inclusion in the NWT Water Strategy program. These partnerships greatly enhance the Dehcho water monitoring program by providing a database that integrates traditional and local knowledge from the communities with western science.

Box 9.4 NWT-Wide Community-Based Water Quality Monitoring Program

During engagement on the development of the Water Strategy, Aboriginal governments, communities and NWT residents expressed keen interest in participating meaningfully in community-based research and monitoring to address their water-related questions and concerns. To address this, the NWT-wide Community-based Water Quality Monitoring Program ('CBM program') was developed. ENR-GNWT has partnered with more than 20 communities, as well as other water partners, to collect a wide variety of water quality data to address community concerns and questions about water quality. Sampling is done at more than 40 sites throughout the NWT that are important to community partners. The program samples for basic water quality parameters and dissolved metals and hydrocarbons. The suite was chosen to address the key concerns raised by community partners, particularly with respect to metals and hydrocarbons that may be linked to upstream development.

Through the program, community partners have received training through workshop modules and on-site. This has resulted in an increase in trained community monitors. Community monitors and ENR-GNWT technicians work together to collect the samples. Currently, collaborative efforts are underway to expand training opportunities to include additional monitoring parameters (e.g., wildlife and fish) and identify longer term employment opportunities for trained monitors. There has been a high degree of interest in the program and the number of communities involved has grown from 12 in 2012 to more than 20 in 2015. The ultimate goal of the CBM program is that, as capacity is built, community partners will take over implementation of the program. One of the challenges with the program has been capacity to analyze and interpret data and get results out to communities and other interested parties in a timely manner. An approach to addressing this challenge has been the development of Mackenzie DataStream, a project by Walter and Duncan Gordon Foundation in partnership with ENR-GNWT, with input from the ASC and water partners. DataStream is an open-source data management system for exploring and sharing water quality data in a manner that is widely accessible to any interested parties.

For more on Mackenzie DataStream, please go to: http://www.mackenzie-datastream.org

9.4.4 Multi-level Linkages

As Bogardi et al. (2012) note, "water security in the 21st century will require better linkage of science and policy, as well as innovative and cross-sectoral initiatives, adaptive management and polycentric governance models that involve all stakeholders" (p. 35). One of the areas where the Water Strategy has been most effective has been linking water partners at multiple levels. All Water Strategy activities are embedded within the broader vision and goals (as outlined above) and all activities are designed to contribute to meeting those goals. Water Strategy initiatives encourage the use of multiple knowledge systems to inform decision-making and stewardship at multiple levels from the local to MRB-wide. Information gathered via different initiatives may scale up or down given the nested nature of decision-making.

Box 9.5 The Significance of the Water Strategy to Achieving Water Security in the NWT, by the Honourable J. Michael Miltenberger, Minister of Environment and Natural Resources (2007–2015)

The challenge faced by the NWT, as we at long last assumed responsibilities for most lands and waters throughout the territory from Canada (which occurred on April 1, 2014), was to know firmly and clearly what we wanted. What was that future we wished to create? Without knowing this, any future could appear to suffice, but not all would protect ecosystems, the northern lifestyles, and our commitments to Aboriginal governments and northern citizens. Water security meant getting our policy house in order and articulating a sustainable integrated vision. For us, this vision is based on three things: first, understanding that integrated water management is primarily about ecosystems, not just water. Second, it means integrating water management between jurisdictions to collaboratively address Mackenzie River Basin needs and third, it means ensuring that local and regional choices – such as water, energy, and food policies and laws – serve the health of the basin. This is what *Northern Voices, Northern Waters* (the Water Strategy) does for NWT: it provides the policy basis for all of NWT's governments – federal, territorial, and Aboriginal – to engage with upstream jurisdictions in a way that achieves these integrated goals towards a sustainable and secure water future.

Similarly, community-based research and monitoring is often undertaken at the local level but scales up to regional, NWT and MRB-wide. While programs such as the NWT-wide community-based water quality monitoring program (see Box 9.4) are geared more specifically to the local level, they are embedded within higher units of analysis (i.e., sub-watershed, nested watershed units). Transboundary water management agreement negotiations are at the basin level but affect perception and decision-making within the NWT (see Box 9.5).

At the Basin-level, one of the core mechanisms for linking jurisdictional mandates and activities is the MRBB. Approaches to water management in upstream jurisdictions provided at MRBB meetings ensure learning at the basin level is considered in water management decision making in the NWT. However, jurisdictional fragmentation can be a challenge (Rosenberg International Forum on Water Policy 2013). The development and signing of bilateral water management agreements will further enhance cooperative efforts towards water management between neighboring jurisdictions and for the MRB as a whole (see Box 9.1).

Water can act as a bridge between diverse water partners – or as a builder of 'community' – in that water in the NWT is a shared resource, and management, use or protection of water is for many groups, a shared goal (see Box 9.5). Linking water partners at multiple levels (individuals, communities, industry, ENGOs, academics, municipal, Aboriginal or territorial government) builds flexible partnerships where water partners can become as involved as they wish in areas of interest to them related to water security. Input from multiple levels strengthens the integrity of implementation activities of the Water Strategy. Participation in Canada-wide and international partnerships ensures this learning can be scaled up nationally and internationally.

9.4.5 Integration

The Water Strategy is holistic in nature. It recognizes that robust, informed decisions come from bringing together information on all parts of the system, multiple knowledges, all of the interests and values of people who use and rely on the waters of the NWT and MRB and all of the people who make decisions related to how water is used and managed. The Water Strategy also approaches water management from the perspective of linking environment, economy and society, with recognition of the need for sustainable development (see Box 9.5).

The Water Strategy advocates for, and adopts, an ecosystem-based approach to management. This approach "requires that those who make decisions which may affect water understand and consider structure, function and processes within ecosystems, as well all values within the watersheds" (GNWT/INAC 2010 p. 12). Watersheds are positioned as the starting point for consideration. Smaller subwatersheds are nested within larger watersheds and within the larger MRB and such linkages must also be considered in decision-making. This requires water partners at all levels to work together, share information and consider interrelations between all parts of the system. If the appropriate partners are not engaged, there is risk that core linkages within the watershed or between watersheds may not be considered, resulting in fragmentation. Geopolitical boundaries can exacerbate these types of issues, which highlights the importance of the MRBB and bilateral water management agreements for promoting and ensuring ecosystem-based management. However, in the NWT and MRB substantial knowledge gaps related to ecosystem processes and health exist, making implementing integrated water management approaches challenging. One of the key functions of partnerships developed under the Water Strategy is to work together to fill these gaps by gathering and utilizing information from multiple knowledge systems and partners.

Though the Water Strategy does not explicitly identify use of an IWRM approach, the intent of the Strategy, through bringing together different mandates, water users, perspectives and levels of decision-making in a collective responsibility for water stewardship, is integrative in nature. Broad framings of water security can complement perspectives on IWRM (Cook and Bakker 2012; Dunn et al. 2012), which can provide a framework for working towards the goal of water security (Bogardi et al. 2012; Global Water Partnership 2012). An IWRM approach is consistent with globally adopted sustainable development agendas, such as in the United Nations *Transforming our world: the 2030 Agenda for Sustainable Development* (United Nations 2015).

In the NWT, there are many parties who are responsible for decisions related to various aspects of water management. Linking groups and mandates is of paramount importance for implementing the Water Strategy to ensure coordinated efforts to protecting water and aquatic ecosystem health. For example, four GNWT departments with roles and responsibilities related to managing drinking water work together to share information and link programs and policies through an interdepartmental working group.

It is important to note that integrated and ecosystem-based approaches are not without criticism or challenges in their implementation. It has been noted that IWRM works well in theory but can be much more elusive and difficult in practice (Cook and Bakker 2012, Biswas 2008; Blomquist and Schlager 2005; Jeffrey and Gearey 2006). The reality is jurisdictional boundaries (at multiple levels), and interests of groups within those boundaries, often cut across watersheds making integration complex and seldom straightforward (Blomquist and Schlager 2005; Dunn et al. 2012). When the watershed is the unit of management, it can be challenging to identify and agree upon boundaries, authority, participation and accountability (Blomquist and Schlager 2005). NWT shares boundary waters with three Canadian provinces and two territories. Municipal, regional and land claim boundaries all cut across various sub-watersheds. Diverse mandates, limited resources and competing interests across vertical and horizontal scales have at times been a challenge in terms of integrating water management. Manifestation of this can be seen in the desire to more clearly identify roles and responsibilities of water partners, as noted earlier, reflecting the complexity in defining authority, participation and accountability.

The ASC guided the collaborative development of the 2016–2020 Action Plan, based on lessons learned from implementation of the Water Strategy between 2010 and 2015, the independent evaluation and engagement with water partners to continue to address challenges related to the use of integrated and ecosystem-based approaches to water management.

9.4.6 Capacity-Building

One of the most critical aspects for collaboratively engaging in activities to promote water security is that interested water partners have the capacity to participate meaningfully in water stewardship. Ensuring water partners who want to be involved in stewardship activities in the NWT are provided with opportunities to learn, participate and collaborate toward implementing policies, programs and making decisions for managing water security is critical to successful implementation of the Water Strategy. The NWT has a small population (approximately 44,000) (NWT Bureau of Statistics 2016), with close to half residing in the capital city of Yellowknife (a little more than 20,000 in 2015; NWT Bureau of Statistics n.d.). The remaining residents of the NWT live among 32 other small communities, spread out over a very large land mass, some quite isolated. In fact, many communities are not accessible by road year round or at all. As such, capacity issues exist for many water partners in the NWT, including ENR-GNWT, and are generally related to limited financial and human capital (though other challenges also exist).

Across communities involved in community-based resource management, there may be variable degrees of financial capital, potentially impacting the capacity to fully address environmental challenges (Bradshaw 2003). This poses an ongoing challenge to successful implementation of collaborative water stewardship. To address financial capacity concerns, water partners have worked together to jointly develop successful funding proposals to support research and monitoring to address community-driven concerns. In 2012, ENR-GNWT produced a water calendar that provided a list of potential funding sources, deadlines for applications to be submitted and tips on how to develop a successful proposal. It was distributed to water partners to increase their capacity to secure financial resources to support stewardship activities. ENR-GNWT and other funders have provided resources for water partners to learn, participate and collaborate on water security-related issues, including source water protection, community-based monitoring and transboundary water management agreement negotiations. This has included community and regional meetings and workshops, as well as ASC meetings and the annual Water Strategy workshop. These events provide opportunities for reciprocal learning among water partners and are important conduits of knowledge from multiple perspectives within the NWT and the MRB. Funders and collaborators have supported ASC members and other NWT residents to attend water meetings in other parts of Canada to learn about what is going on elsewhere to potentially apply and adapt ideas and best practices relevant in the NWT.

One of the biggest challenges in building capacity to undertake work related to the Water Strategy is human resources capacity. This appears to be a common challenge in community-based initiatives (Bradshaw 2003; Sharpe and Conrad 2006). Addressing this is an ongoing challenge and requires creative and strategic thinking. Targeted training, capacity-building initiatives and support are important (Bradshaw 2003; Sharpe and Conrad 2006; Conrad and Hilchey 2011). To this end, water partners continually look for ways to increase training and learning opportunities and build capacity in Water Strategy activities. As noted above, water partners will often work together to access funding to undertake projects, which increasingly involves allocation of financial capital to support building of human resources capacity in partner organizations, with particular emphasis on providing support to community partners to undertake activities. Some recent Water Strategy initiatives have, or will, employ or support community coordinators to fulfill project implementation needs, and ensure the interests of the communities are served and capacity remains within communities for further participation in water security-related activities. This shift has resulted from discussions about the need to support meaningful and sustained capacity at the community level.

Youth are encouraged to participate in learning at meetings and public events. There have been youth projects specifically designed to elicit interest on water security issues and build capacity for their enhanced participation over the long term. Monitors taking part in the ENR-GNWT supported community-based monitoring program have been provided with technical training at workshops and on site while monitoring (see Box 9.4). This has resulted in an increase in trained water quality monitors across the NWT and has continued to grow interest in community-based water monitoring.

Though the Water Strategy has resulted in increased capacity across water partners, there is certainly more work to be done. Building strong capacity to meaningfully participate in water stewardship activities across the NWT is an ongoing endeavour. Changing fiscal outlooks, high degree of personnel turnover and shifting priorities (at multiple levels) can all impact financial and human resource capacity and the ability to successfully undertake water stewardship activities. As such, it remains one of the most critical areas of focus for all water partners and new approaches for targeted training and capacity-building are continually discussed and explored. This is perhaps an area where the NWT could take cues from successful community development and capacity-building initiatives internationally. Such international engagement was one motivation to participate in the previously mentioned Tracking Change project and to learn from partners dealing with similar experiences and challenges in the Amazon and Mekong River Basins.

9.5 Looking Forward: The NWT Water Strategy – Supporting Communities of Practice?

As evident throughout this chapter, one of the strengths of the Water Strategy is its approach to collaboration – how it builds upon existing cooperation and partnerships and how it can continue to foster space to bring people with shared values and interests together. The collaborative nature and partnership aspects of the Water Strategy can point to thinking about, or framing it within, the context of 'communities of practice'. Communities of practice are defined by Wenger-Trayner & Wenger-Trayner (2015) as entities "formed by people who engage in a process of collective learning in a shared domain of human endeavour", and that such groups "share a concern or a passion for something they do and learn how to do it better as they interact regularly" (p. 1). Thus, key to communities of practice is joint learning through regularized interactions and shared practice (Wenger-Trayner & Wenger-Trayner 2015; Pahl-Wostl et al. 2007); though Wenger-Trayner & Wenger-Trayner (2015) do note that learning could be either an intentional or incidental result of engagement in the process.

Thus, we can see parallels between the concept of communities of practice and the elements of collaboration, participation, partnerships and learning embedded within the Water Strategy, that are critical to achieving water security. The concept of communities of practice could have potential explanatory power in understanding the function and operation of the Water Strategy, as well as how groups form, operate and share resources when working towards water security. As Bradshaw (2003) notes, community can be conceptualized in multiple ways. As evident in this chapter, 'community' – and the ways water can bring members of a community together towards common goals – can be approached as a broad and fluid concept when thinking about how people come together to address issues of water (in)security.

It would be remiss to not reflect on some of challenges associated with building or maintaining communities of practice. While a fulsome review of the communities of practice literature is beyond the scope of this chapter, for detailed discussion on key 'ingredients' and examples of successful and unsuccessful communities of practice, readers are directed to Probst and Borzillo (2008). One of the key caveats is that building effective communities of practice generally relies upon substantial investments of time, energy and resources and development of trust, in order to foster reciprocal learning, shared understanding and undertake joint action (Cundill et al. 2015; Roberts 2006).

Given the challenges noted above with financial and human resources, capacity challenges and geographic distance in the NWT, ensuring regularized interactions necessary to build communities of practice is no small task. Indeed, at times, this has posed difficulty in building partnerships to undertake water stewardship. Ongoing commitment to the Water Strategy and to partnerships, meaningful and respectful approaches to collaboration and investments in long-term relationships must continue to be at the forefront of water stewardship activities in the NWT.

As such, there is value in future research that examines how communities of practice come together and function in the NWT (and what the potential pitfalls may be) and how those communities of practice may be embedded within the Water Strategy. Drawing on the communities of practice literature may help to elucidate ways forward for dealing with some of the complex issues and challenges the NWT may face in terms of water security and potentially point to how such collectives may be nurtured (*sensu* Cundill et al. 2015). Recent discussion around 'transdisciplinary communities of practice' in pursuit of sustainability (as introduced by Cundill et al. 2015) may be particularly illuminating in this regard. How the Water Strategy functions to support or create communities of practice – or whether the Water Strategy could be seen as a community of practice in its own right – represents an interesting avenue of future investigation.

9.6 Conclusion

In this chapter, we have explored the Water Strategy in the context of how it contributes to and promotes water security for NWT residents. Though grounded in the experience of the NWT, the lessons derived from the creation and implementation of the Water Strategy will likely resonate with national and international practitioners dealing with similar challenges – whether they are working with diverse actors and stakeholder groups, dealing with multiple knowledge systems or trying to work collaboratively to manage transboundary waters (Boxes 9.6 and 9.7).

Box 9.6 Lessons from the Water Strategy to Support Water Security Nationally, by Merrell-Ann Phare, Executive Director, Centre for Indigenous Environmental Resources

Water security cannot be achieved on a national scale unless all citizens and their governments come together and agree on a path towards that security. This type of cooperation seems insurmountable in the complexity of today's world and the multiple governance systems that operate at numerous levels to manage human actions regarding water. The NWT chose to rely upon the complexity of the governance systems within its system; those arising from land claims and treaties in addition to the federal and territorial government jurisdictions. These multiple governments are a source of strength to NWT, and within Canada, and investing heavily in comprehensive processes built on their collaboration not only strengthens those relationships – which was needed to build the Water Strategy – but also means that the commitment to stay true to the vision is strong. This approach can be, should be, and indeed must be used by the rest of Canada to build strong foundations throughout our basins nation-wide. Water security for Canada will only be built on strong relationships both between the nations within our nation.

Box 9.7 The Water Strategy in an International Context, by Bob Sandford, EPCOR Chair, Water Security, United Nations University, Institute for Water, Environment & Health

The national and international significance of *the Northwest Territories' Northern Voices, Northern Waters: NWT Water Stewardship Strategy* derives from the patient and inclusive manner in which it was crafted and implemented and the on-going engagement process through which its effectiveness will continue to be thoughtfully monitored and evaluated. In that it respects traditional knowledge and ways of life within the context of basin-wide aquatic ecosystem health represents a landmark in integrated watershed management. Its example demonstrates that in fact there are no real jurisdictional, legislative, constitutional or political obstacles to the creation of a sustainable future. All it takes is political will and persistent leadership. It is no longer possible anywhere to say it can't be done. The Northwest Territories just did it.

Management to ensure safe, abundant waters for people and ecosystems has bearing beyond the NWT. As noted in the introduction to this chapter, much of the NWT's water is part of the MRB, which has major impacts globally. Therefore, maintaining water security in the NWT can support global processes affecting others outside of the NWT and even Canada. What the NWT does to manage its waters matters on the world stage. In offering insights from our experiences, it is our hope that what we have learned and accomplished both through our successes and our challenges can be useful to others working to advance and achieve water security within complex social-ecological systems.

Acknowledgments The authors wish to gratefully acknowledge all those involved in Water Strategy development and implementation in the NWT, including the Aboriginal Steering Committee, and all water partners, as this chapter is a reflection of the work of many. Thank you to George Low, Merrell-Ann Phare, J. Michael Miltenberger and Bob Sandford for contributing text boxes and analysis of the role of the Water Strategy. The authors also wish to thank the editors, an anonymous reviewer and Bob Sandford for their helpful and insightful reviews of the chapter.

References

Aboriginal Affairs and Northern Development Canada. (2010). *Water today, water quality and quantity in the NWT.* Aboriginal Affairs and Northern Development Canada. Available from: http://www.aadnc-aandc.gc.ca/DAM/DAM-INTER-NWT/STAGING/texte-text/ntr_pubs_wt10_1330704074233_eng.pdf. [December 19, 2015]

Adams, M. S., Carpenter, J., Housty, J. A., Neasloss, D., Paquet, P. C., Service, C, et al. (2014). Toward increased engagement between academic and indigenous partners in ecological research. *Ecology and Society, 19*(3), 5.

Allan, J. A. (2013). Food-water security: Beyond water resources and the water sector. In B. Lankford, K. Bakker, M. Zeitoun, & D. Conway (Eds.), *Water security: Principles, perspectives and practices* (pp. 321–335). New York: Routledge.

Armitage, D., Berkes, F., Dale, A., Kocho-Schellenberg, E., & Patton, E. (2011). Co-management and the co-production of knowledge: Learning to adapt in Canada's Arctic. *Global Environmental Change, 21*(3), 995–1004.

Armitage, D., de Loë, R. C., Morris, M., Edwards, T. W., Gerlak, A. K., Hall, R., et al. (2015). Science-policy processes for transboundary water governance. *Ambio, 44*(5), 353–366.

Berkes, F. (2008). *Sacred ecology* (2nd ed.). New York: Routledge.

Berkes, F., Colding, J., & Folke, C. (2000). Rediscovery of traditional ecological knowledge as adaptive management. *Ecological Applications, 10*(5), 1251–1262.

Biswas, A. K. (2008). Integrated water resources management: Is it working? *International Journal of Water Resources Development, 24*(1), 5–22.

Blomquist, W., & Schlager, E. (2005). Political pitfalls of integrated watershed management. *Society and Natural Resources, 18*(2), 101–117.

Bogardi, J. J., Dudgeon, D., Lawford, R., Flinkerbusch, E., Meyn, A., Pahl-Wostl, C., et al. (2012). Water security for a planet under pressure: Interconnected challenges of a changing world call for sustainable solutions. *Current Opinion in Environmental Sustainability, 4*(1), 35–43.

Bradshaw, B. (2003). Questioning the credibility and capacity of community-based resource management. *The Canadian Geographer/Le Géographe Canadien, 47*(2), 137–150.

Conrad, C. C., & Hilchey, K. G. (2011). A review of citizen science and community-based environmental monitoring: Issues and opportunities. *Environmental Monitoring and Assessment, 176*(1–4), 273–291.

Cook, C., & Bakker, K. (2012). Water security: Debating an emerging paradigm. *Global Environmental Change, 22*(1), 94–102.

Cundill, G., Roux, D. J., & Parker, J. N. (2015). Nurturing communities of practice for transdisciplinary research. *Ecology & Society, 20*(2), 22.

de Loë, R., Varghese, J., Ferreyra, C., & Kreutzwiser, R.D. (2007). *Water allocation and water security in Canada: Initiating a policy dialogue for the 21st century*, Guelph Water Management

Group, University of Guelph. Available from: http://www.wpgg.ca/sites/default/files/1-de_Loe_et_al_2007_Final_Report.pdf. [December 2015]
Dunn, G., Cook, C., Bakker, K. & Allen, D. (2012). *Water security guidance document*, UBC Program on Water Governance. Available from: https://watergovernance.ca/projects/water-security/water-security-manual-2/. [May 29, 2015]
Froggatt, A. (2013). The water-energy nexus: Meeting growing demand in a resource-constrained world. In B. Lankford, K. Bakker, M. Zeitoun, & D. Conway (Eds.), *Water security: Principles, perspectives and practices* (pp. 115–129). New York: Routledge.
Global Water Partnership. (2012). *Increasing water security – A development imperative*, Global Water Partnership. Available from: http://www.gwp.org/Global/About%20GWP/Publications/Perspectives%20Paper_Water%20Security_final.pdf. [May 29, 2015]
GNWT. (2005). *GNWT Traditional Knowledge Policy 53.03*, Government of the Northwest Territories. Available from: http://www.enr.gov.nt.ca/sites/default/files/documents/53_03_traditional_knowledge_policy.pdf. [May 29, 2015]
GNWT/INAC. (2010). *Northern voices, northern waters: NWT water stewardship strategy*. Yellowknife: GNWT.
Grey, D., & Sadoff, C. W. (2007). Sink or swim? Water security for growth and development. *Water Policy, 9*(6), 545–571.
Jeffrey, P., & Gearey, M. (2006). Integrated water resources management: Lost on the road from ambition to realisation? *Water Science & Technology, 53*(1), 1–8.
Ludi, E. (2009). *Climate change, water and food security*. Overseas Development Institute Background Note. Available from: http://www.odi.org/sites/odi.org.uk/files/odi-assets/publications-opinion-files/4116.pdf. [May 29, 2015]
Mackenzie Global Energy and Water Cycle Experiment (GEWEX) Study. (2004). Available from: http://www.usask.ca/geography/MAGS/. [May 29, 2015]
Mitchell, B. (2006). IWRM in practice: Lessons from Canadian experiences. *Journal of Contemporary Water Research & Education, 135*(1), 51–55.
MRBB. (2003). *Mackenzie River Basin state of the aquatic ecosystem report 2003*. Yellowknife: Mackenzie River Basin Board, NWT.
Norman, E.S., Bakker, K., Cook, C., Dunn, G. & Allen, D.. (2010). *Water security: A primer*, UBC Program on Water Governance. Available from: http://watergovernance.ca/wp-content/uploads/2010/04/WaterSecurityPrimer20101.pdf. [December 15, 2015]
NWT Bureau of Statistics. (2016). *Population estimates*. Available from: http://www.statsnwt.ca/population/population-estimates/. [January 14, 2016]
NWT Bureau of Statistics. (n.d.). *Yellowknife–Statistical Profile*. Available from http://www.statsnwt.ca/community-data/Profile-PDF/Yellowknife.pdf. [January 17, 2016}.
NWT Water Stewardship. (n.d.). *Water is life*. Available from: http://www.nwtwaterstewardship.ca/water_is_life. [May 29, 2015]
Office of the High Commissioner for Human Rights, UN Habitat & World Health Organization. (2010). *The human right to water: Fact Sheet No. 35*. Available from http://www.ohchr.org/Documents/Publications/FactSheet35en.pdf. [December 19, 2015]
Pahl-Wostl, C., Craps, M., Dewulf, A., Mostert, E., Tabara, D., & Taillieu, T. (2007), Social learning and water resources management, *Ecology and Society*, *12*(2).
Palmer, M., Higgins, J. & Gah, E. (2012). *A freshwater classification of the Mackenzie River Basin*. Northwest Territories Protected Areas Strategy. Available from: http://www.nwtpas.ca/documents/document-2013-freshwaterclassification.pdf. [May 29, 2015]
Plummer, R., Baird, J., Moore, M.-L., Brandes, O., Imhof, J., & Krievins, K. (2014). Governance of aquatic systems: What attributes and practices promote resilience? *International Journal of Water Governance, 2*(4), 1–18.
Pohl, C., Rist, S., Zimmermann, A., Fry, P., Gurung, G. S., Schneider, F., et al. (2010). Researchers' roles in knowledge co-production: Experience from sustainability research in Kenya, Switzerland, Bolivia and Nepal. *Science and Public Policy, 37*(4), 267–281.

Probst, G., & Borzillo, S. (2008). Why communities of practice succeed and why they fail. *European Management Journal, 26*(5), 335–347.

Roberts, J. (2006). Limits to communities of practice. *Journal of Management Studies, 43*(3), 623–639.

Rosenberg International Forum on Water Policy. (2013). *Rosenberg International Forum: the Mackenzie River Basin, Report of the Rosenberg International Forum's Workshop on Transboundary Relations in the Mackenzie River Basin*, Rosenberg International Forum on Water Policy. Available from: http://gordonfoundation.ca/sites/default/files/publications/Rosenberg%20Final%20-%20WEB.pdf. [December 15, 2015]

Sharpe, A., & Conrad, C. (2006). Community based ecological monitoring in Nova Scotia: Challenges and opportunities. *Environmental Monitoring and Assessment, 113*(1–3), 395–409.

Tengö, M., Brondizio, E. S., Elmqvist, T., Malmer, P., & Spierenburg, M. (2014). Connecting diverse knowledge systems for enhanced ecosystem governance: The multiple evidence base approach. *Ambio, 43*(5), 579–591.

United Nations. (2015). *Transforming our world: The 2030 Agenda for Sustainable Development.* Available from: https://sustainabledevelopment.un.org/content/documents/21252030%20Agenda%20for%20Sustainable%20Development%20web.pdf. [December 19, 2015]

UN-Water. (2013). *Water security and the global water agenda: A UN-Water analytical brief.* United Nations University Institute for Water, Environment and Health. Available from: http://www.unwater.org/downloads/watersecurity_analyticalbrief.pdf. [December 15, 2015]

Wenger-Trayner, E. & Wenger-Trayner, B. (2015). *Communities of practice: A brief introduction*, Wenger-Trayner. Available from: http://wenger-trayner.com/wp-content/uploads/2015/04/07-Brief-introduction-to-communities-of-practice.pdf. [May 29, 2015]

World Economic Forum. (2011). Introduction: The water-food-climate-energy nexus: A facts and figures overview. In World Economic Forum (Ed.), *Water security: The water-food-climate-energy nexus* (pp.1–16). Washington, DC: Island Press.

Chapter 10
Pathways to a Water Secure Community

Corinne J. Schuster-Wallace and Sarah E. Dickson

Abstract Community-based water security, defined herein as "the sustainable access to affordable and reliable quantities of water of suitable quality to enable all persons to lead healthy, dignified, and productive lives, including neighbours and future users", has not been paid a significant amount of attention to date. However, in light of current global access to drinking water and sanitation facilities, wastewater treatment coverage, the importance of water for food, energy and industry, and the impacts of climate change on the hydrological cycle, water (in)security is extremely important at the local scale and a potential threat to (de)development. While community water security can be difficult to incorporate into the water security continuum due to differences in scale, water secure communities are a building block for sustainable watersheds. After identifying the different aspects of water security at the community level, it is concluded that a comprehensive, systems approach coupled with capacity for sustainable local change is key for sustained and sustaining community water security.

Keywords Water security • Rural, remote, marginalised • Community • Integration • Pathways

10.1 Introduction

Water security, particularly in national and transboundary basin contexts, has been the subject of research, dialogues, and political agreements for decades. Community water security, in contrast, is an emerging topic that has not historically been considered as part of the continuum of scale for water security. However, it is clear that water (in)security is important at the local scale, particularly in light of global environmental change processes and the post-2015 sustainable development agenda. Individual communities shape patterns of water use and impact water quality and

C.J. Schuster-Wallace (✉)
Independent Water-Health Expert, McMaster University, Hamilton, ON, Canada
e-mail: corinne.schusterwallace@gmail.com

S.E. Dickson
Department of Civil Engineering, McMaster University, Hamilton, ON, Canada

D. Devlaeminck et al. (eds.), *The Human Face of Water Security*, Water Security in a New World, DOI 10.1007/978-3-319-50161-1_10

quantity that, in turn, contributes to water (in)security of larger and larger nested watersheds. Community water insecurity also has larger socio-economic and environmental impacts, a ripple effect creating burdens on health care systems and national economies through poor health and wellbeing. Indeed, as articulated by Bradley and Bartram (2013), "provision of water for human domestic consumption can be viewed as a fundamental example of water security" (p. 2). In developed societies, water insecurity, especially manifested through natural disasters, could actually reverse standards of development (i.e. de-development), in a manner similar to that previously identified in conflict situations (e.g. Roy 1999).

Community water security can be difficult to incorporate into the larger water security dialogue because communities do not always have resources available to engage and community problems manifest as local problems with local impact. However, building blocks for water security, such as water demand, local economic activity, and an educated and productive workforce start at this community level. As such, this chapter explores the various components of water security at the community level, the barriers and challenges faced by communities wanting to become water secure, and introduces tools that support an understanding of community water security status and thus informs decision-making, before defining pathways and immediate next steps to achieve community water security.

Communities can be defined as a group of people living in proximity to each other, a group of people with similar social and/or cultural perspectives, a group of people sharing the same space e.g. a hospital community, or place of work. Within the context of a water secure community, we chose the definition of proximal living. At one end of the spectrum in this definition are large urban centres. Water security in this setting is unique and requires emphasis on closing the water loop and maximizing water efficiency within the system. In this chapter, however, the focus is placed on rural, remote and otherwise marginalized communities (RRMs). These communities tend to be least served when it comes to infrastructure and social services. This is attributable to several factors, including difficult physical access, lack of existing infrastructure, small populations and therefore low tax or capital asset bases, lack of local capacity and/or retention difficulties, and low population density which hinders benefits from economies of scope and scale. The bottom line is that extension of traditional centralized services is often much more expensive per connection than urban centres and therefore largely unattainable. While all elements of the framework can be applied to urban settings, the mechanisms by which these are achieved and the responsibility for them may vary considerably. For example, food security in large urban centres is likely to be externally determined through domestic food production and food imports. Jurisdiction can become far more complicated and require intersectoral mechanisms to ensure sustainable security with respect to water. Moreover, while RRMs tend to be disadvantaged in comparison to their urban centre counterparts, huge inequities and vulnerabilities can exist within urban settings, especially for people living in informal settlements.

10.2 Community Water Security

Water security depends on and impacts many physical, social, economic, political, and institutional factors both in the community and the broader context in which they are found. It has been defined in many different ways and at many different scales (from global to community). Cook and Bakker (2012) documented the growing use of the term 'water security' across many disciplines and identify five cross-cutting themes: availability; water-related hazards; vulnerability to those hazards; human needs; and, environmental sustainability. They argue that these themes must be brought together in an holistic conceptual framework, incorporating environmental, social, and economic factors, and conclude that applied in this manner, water security holds promise as an emerging water management strategy. Falkenmark (2001) discusses the need to consider water as a driver for socio-economic development. Grey and Sadoff (2007) also consider water as a socio-economic driver, but at the much smaller scale of livelihood, to which water is essential for poverty reduction; thus goals of water security cannot be realized without addressing water for livelihoods.

When considering the concept of scale in the definition of water security, Lundqvist et al. (2003) emphasizes the need for both a basin-level authority and national policy to achieve urban water security in developing countries. This is supported by the World Water Council vision (Cosgrove and Rijsberman 2000), which further advocates for the inclusion of ecosystem services in the definition of water security. They note that we are only beginning to understand the value of these services, which include flood control, irrigation, industry, recreation and waterway transportation. The Asian Development Bank (2013) identified five key dimensions to water security at the national scale, including: household, urban, environmental, resilience to water-related disasters, and economic. At the other end of this scale, Siwar and Ahmed (2014) presented three elements of domestic-level water security: safety, access, and affordability, required for every person to lead a clean, healthy, productive life while ensuring the natural environment is maintained. Bogardi et al. (2012) note that water connects socio-ecological, economic, and geophysical systems at multiple scales and therefore emphasize the need to incorporate the concept of the global water system into the definition of water security. As we are constantly perturbing this system, we create the potential for water insecurity as evidenced by the impacts of climate change. Indeed, Calow et al. (2010) emphasize that water security mapping at the community level can significantly increase community resilience to climate variability through targeting resources to vulnerable areas.

Grey and Sadoff (2007) also incorporate climate change into their definition of water security through the concept of risk, noting that it is not only the absence of water that poses a risk, but that the presence of water can also pose a risk to people, economy and the environment through, for example, disasters and the transmission of disease. More recently, others have incorporated risk as a central piece of water security. OECD (2013) advocates for a risk-based approach to water security, in which the level of acceptable risk, in terms of the likelihood that an event will occur and its impact if it does, is weighed against the benefits of improving water security. The report argues that this approach is both proactive and flexible. Grey et al. (2013) describe water security as 'a tolerable level of water-related risk to society'. Bradley and Bartram (2013), however, argue that these risk-focused definitions are most appropriate for populations who have at least some water, since they emphasize the risk of it being taken away. They argue that a more comprehensive definition of water security incorporates both provision and risk, and suggest that provision must go beyond the household level and address maintenance and equity of access in addition to upfront capital costs.

Incorporating and adapting these concepts for community-scale water security results in a working definition that forms the cornerstone of this chapter, i.e., community water security is "the sustainable access to affordable and reliable quantities of water of suitable quality to enable all persons to lead healthy, dignified, and productive lives, including neighbours and future users" (Dickson et al. 2016). Within the context of this definition, we propose a community water security framework incorporating nine dimensions. These dimensions impact upon or are impacted by water (in)security, particularly in RRM communities (Fig. 10.1). The nine dimensions are water resources, water infrastructure, sanitation/wastewater infrastructure, access and equity, community capacity and capital, health and wellbeing, food security, economic activity, and environment. These dimensions become recurring themes in the following subsections, which demonstrate how they are embedded in and weave through community water, environmental, food, economic and health security.

The application of the community water security framework shown in Fig. 10.1 requires an integrated systems approach for sustainable development – social development, environmental integrity, and economic growth. In this context, community water security builds upon the notion of domestic or personal water security, which is explored by Bradley and Bartram (2013). Domestic water security is recognized as a critical element in community water security in our conceptualization (Fig. 10.1), along with food, livelihood, and environmental security. In this manner, community water security is implicitly tied to the health of community members; health not simply measured as the presence or absence of disease, but a reflection on physical capacity, access to personal resources, and access to social resources (WHO 1986).

Health: a resource for everyday life, not the objective of living (WHO 1986).

Fig. 10.1 A framework for RRM community water security

10.2.1 Access to Water Resources

Molle and Mollinga (2003), identify five categories of water use, noting that each one has a different impact on human well-being: drinking water, domestic water, food security needs, economic production, and environmental needs. Drinking water is the least flexible; it is required to sustain life, while economic production, albeit important for long-term health and wellbeing, is secondary. Water use depends on availability. However, it has been suggested that the number of countries facing water scarcity by 2030 will increase dramatically, with an estimated 47% of the population living in water-stressed regions by that time (UNESCO 2009). It is estimated that currently 166 million people in 18 countries are affected by water scarcity, and an additional 270 million live in 11 countries which are identified as water stressed (Jury and Vaux 2007). Understanding the cause of scarcity is central to the ability to combat it. There are five causes of water scarcity: physical scarcity, economic scarcity, managerial scarcity, institutional scarcity and political scarcity (Molle and Mollinga 2003). Physical scarcity refers to the absence of water in nature i.e. water resources. Economic scarcity occurs when the resources required to access water, either human or financial, are lacking. Managerial scarcity arises from inadequate management or maintenance of water infrastructure. Institutional scarcity is a more subtle form of managerial scarcity (Molle and Mollinga 2003) in which general supply and demand imbalances are not anticipated and therefore not mitigated. Political scarcity occurs when people are prevented from accessing water. These five causes of water scarcity link to the water resources, water infrastructure, community and social capacity, and access and equity dimensions of the water security framework presented in Fig. 10.1. Considering the five categories of use together with the five causes of scarcity, Molle and Mollinga (2003) conceive of 25 cases of different kinds of water scarcity associated with different underlying causes, which emphasizes the complexity of the scarcity issue.

Scarcity may vary temporally, as inter- and intra-annual rainfall is often variable, and is becoming more so as a consequence of climate change. This highlights the potential for institutional scarcity to increase, as it is difficult to anticipate and plan for rainfall patterns that are consistently changing. This is compounded by the fact that current prediction models are based on the assumption that historical patterns are not dissimilar to future ones (hydrologic stationarity). Climate change has rendered hydrologic stationarity a misconception (Milly et al. 2008). Despite climate change and associated uncertainties, rainwater harvesting (RWH) is recognized as a necessary tool for addressing water scarcity, particularly where economic, institutional, and political reasons are the underlying causes of scarcity. RWH alone, however, is insufficient to bridge the dry seasons in regions where they exist, as it is impractical to incorporate the roof and storage size combinations that would be necessary to achieve this (El Ganzouri et al. 2015). Thus, resilience must be built into water provisioning plans through incorporating backup sources to minimize the risk of scarcity, and ultimately achieve security within the context of the water resources dimension of the framework.

From a security perspective, water quality is as essential as quantity, and is embedded in the accessibility, environment, and sanitation dimensions of the water security framework in Fig. 10.1. Whereas, at present, if water infrastructure is in place, treated water is used for all purposes, expenditure of unnecessary resources is not sustainable. Within a community water security framework, quality of water must be fit to the purpose with guidelines and standards to protect health. The required quality is dependent upon the intended use, with the highest quality required for consumption. Water available in sufficient quantities, but of insufficient quality for the purpose, is not truly accessible.

A major component of achieving or maintaining water of an acceptable quality is adequate sanitation and wastewater treatment. Faecal matter contains microorganisms, many of which are pathogenic (i.e. harmful to human health) and may become waterborne when faeces and water come into contact. The mechanisms and infection routes of water-related diseases are discussed further in the next subsection of this chapter.

Ecosystem services are also imperative to achieving and protecting water quality. The hydrologic cycle acts as a natural treatment process; contaminants are left behind when water vaporizes and are filtered as it migrates through soil. Plants uptake contaminants through their root systems and return clean water to the atmosphere through the transpiration process. Ecosystem services are also imperative to quantity of water resources. The natural environment provides storage mechanisms both above and below ground in the form of, for example, snowpacks, surface water bodies, and aquifers. The hydrologic cycle continually replenishes these reservoirs as they are depleted.

Community-scale access to these water sources is an important dimension in the water security framework presented in Fig. 10.1, and can be a significant challenge in RRM communities. Physical access includes considerations such as proximity and safety. The route from a home to the nearest source often presents numerous barriers, and the distance can vary seasonally as some routes become treacherous during rainy seasons. The time it takes to traverse this route must also be considered, and cannot be excessive, as it is time that can be spent on other productive activities such as obtaining an education or earning an income. Physical ability also becomes an issue as the distance between a home and the water source increases, particularly for the young, elderly, and ill. Safety, from both animals and people, can be a barrier with risk increasing as the distance between the home and source increases.

A person may be prevented from accessing a source due to social barriers, which incorporate both equity and affordability. For example, people may not have claim of ownership over a water source, or may be stigmatized or marginalized within a community, preventing access to certain water sources even though these may be physically accessible. Some may choose not to access a water source due to cultural or personal beliefs, putting themselves at risk. For example, the taste of chlorine in treated water supplies can lead to some groups preferring non-treated sources. Access is sometimes limited when the amount of water that can be supplied by a source is insufficient for all users; those belonging to lower castes or classes, or other marginalized groups who do not hold power, are often granted limited and insufficient access.

10.2.2 Water, Sanitation and Hygiene

Water, sanitation, and hygiene, or WaSH as it is commonly referred to, is an essential pillar of community water security, impacting upon or impacted by several dimensions as per Fig. 10.1. Specifically, inadequate WaSH affects water resources, environment, access and equity, and health and wellbeing. Lack of sanitation and treatment of waste results in pollution of surface water and some ground water aquifers while adequate WaSH reduces exposure to water-related diseases, and access to water and sanitation facilities are important for practicing proper hygiene behaviours.

Human and animal faeces contain bacteria, viruses, and protozoa that are pathogenic (e.g. *Escherichia coli*, *Norovirus*, *Giardia lamblia,* and *Cryptosporidium parvum*). If a community practices open defecation, lacks adequate sanitation, or does not treat their waste, these pathogens end up on or in the soil, if not directly in water bodies. When it rains, pathogens can be transported into surface waters or into shallow aquifers, particularly if there are shallow and/or poorly constructed wells in the vicinity. In the same way, animal faeces deposited near waterways can be washed into the source water. Once in the water, unless treated before use or drinking, these pathogens enter our bodies causing disease. There are several transmission pathways for these pathogens to enter the digestive system including direct consumption of contaminated water, on food products washed with contaminated water, or through bathing in contaminated water. Other water-related diseases, such as malaria and dengue are mosquito borne, with breeding occurring in still or slow moving water, from water in old tires to lakes and ponds. Schistosomiasis is another water-related disease that is transmitted by snails.

Many waterborne diseases present with diarrhoea which leads to dehydration and, if not treated, death, particularly in children under five years of age. Approximately 600,000 children die from diarrhoea every year (Clasen et al. 2014) and many more are sick. This sickness has ramifications not only for the healthcare system, which can be overwhelmed by patients presenting with these preventable infectious diseases, but for education and economic activity. Children and adults who are sick will either stay at home or not be as productive in school or the workplace. Women, as traditional caregivers in the home, lose out on productive labour as a result of having to care for sick family members. As traditional water collectors, their burden also increases through increased need for water to clean up after those who are ill, especially as a result of gastrointestinal diseases. Children miss out on their education. More importantly, chronic diarrhoea in young children prevents absorption of critical nutrients, which can lead to cognitive and physical impairment e.g. stunting (Mullin and Incerpi 2010; Prendergast and Kelly 2012; Smith and Hadad 2014). Not only do these prevent individuals from achieving their maximum potential, costing national economies in lost productivity, but physical impairment in females can lead to complications in labour and delivery, increasing rates of maternal and newborn mortality. Hutton (2012) estimates that these productivity losses cost countries up to US$260 billion annually; conversely, the average global investment in WaSH returns $5.5 per $1 invested.

A similar burden is placed on individuals, health care systems, and the economy from other water-related diseases. For example, poor management of water leads to mosquito breeding. Globally, vector borne diseases contribute an estimated 17% of the infectious disease burden (WHO 2014a). More specifically, WHO estimates that there are between 50 and 100 million cases of dengue annually (WHO 2014b) and approximately 200 million cases of malaria each year (WHO 2014c). People seeking care place a large and unnecessary burden on health care systems, especially when sleeping under nets can virtually eliminate the risk of contracting malaria, as the mosquitoes responsible for transmitting the disease are night biters. Dengue is more difficult to prevent, as the mosquito hosts are out during the day. Schistosomiasis is another water-related disease with a high burden of illness (Prüss Ustün et al. 2008) which contributes to sick days and presenteeism (participation at lower levels of productivity than when healthy).

Given the links between water, human waste, and people, water security from a disease perspective can only be achieved through the integration of potable drinking water, sanitation and faecal sludge management, and personal hygiene (WaSH). More importantly, this has to be practiced by all community members in order to avoid person-to-person transmission and local environmental exposure to pathogens. With 2.4 billion people without access to improved sanitation[1] and over one billion people currently practicing open defecation (WHO and UNICEF 2015), this is clearly an ongoing problem despite significant progress to date. Moreover, improved drinking water sources[2] do not necessarily equate with potable water. Indeed, it is estimated that, even though only 663 million people do not have access to improved drinking water sources (WHO and UNICEF 2015), almost two billion do not have access to water of sufficient quality to protect health (Onda et al. 2012). Hygiene is key in ensuring that pathogens are not spread. Hands can transfer pathogens to mouths, food and surfaces, particularly if not washed after defecating. Handwashing with soap is an effective way of removing germs and preventing person-to-person transfer of disease as well as preventing water-washed diseases such as trachoma, which can cause blindness.

In a similar manner, it is not enough to focus solely on household access. In order for a community to be WaSH secure, WaSH must be addressed in communal meeting places, health care centres, and schools. The latter are critical, as children and the sick are two of the most vulnerable groups when it comes to water-related disease outcomes. Not only are they critical from a vulnerable population perspective, but they are critical from a gender perspective. Lack of sanitation in schools places an additional burden on women and girls post-puberty, who can miss up to a week of school each month due to a lack of adequate menstrual hygiene facilities. Lack of WaSH facilities and water-related diseases during pregnancy and childbirth pose risks to both the mother and baby (Schuster-Wallace and Watt 2015). Examples include lack of nutrients resulting in physical and cognitive defects at birth, increased risk of haemorrhaging during childbirth as a side effect of anaemia from malaria infection and post-birth sepsis (infection). Hygiene is essential in health care settings for infection control and is almost impossible in the absence of the other elements of WaSH, namely water and sanitation, as seen in the recent Ebola outbreak in West Africa (UNICEF 2014).

[1] Improved sanitation facilities separate people from contact with their faeces, with the lowest improved facility being the ventilated improved pit latrine (VIP) (WHO and UNICEF 2015).

[2] Improved drinking water sources are protected from contamination in some way e.g. a cover on a dug well or a wall around a spring (WHO and UNICEF 2015).

10.2.3 Water for Food Security

Nutrition is an important element of health and development, impacted not only by water-related diseases, but by access to sufficient quantity and variety of foods. In many RRMs in low and middle income countries (LMICs), food production occurs at the subsistence level (e.g. kitchen gardens), or through small farm holders. Given that more than one billion people live in dryland regions (UNESCO 2009), and that crop productivity under irrigation is at least double that of rain fed agriculture (World Bank 2008), water is an important element for local food security and water insecurity will have an impact upon food security and therefore community health. This is highlighted in the fact that water stress is a reality for just over one third of the rural population in LMICs (World Bank 2008).

Agriculture is also an important economic sector and source of livelihood for many rural inhabitants in LMICs, many of whom live below the poverty line. According to the World Bank (2008), approximately 75% of LMIC populations live in rural areas and of those, over two billion live on less than $2 per day, while 2.5 billion are involved in agricultural activity. Human waste is a natural fertilizer that increases size of individual plants and produce harvested as well as supports crop diversity. However, untreated human excrement is a source of pathogens which can contaminate plants. Thus, while human waste has the potential to increase productivity, thereby transitioning from subsistence to livelihood agriculture, human waste must be treated before use.

10.2.4 Water for Economic Growth

In addition to the impacts of WaSH insecurity upon economic growth, namely lost productivity and individual potential, and the need for water in agriculture, water plays a key role in other economic sectors, such as energy, processing, and manufacturing. Small scale value-added processing can be a source of community entrepreneurship and economic growth in RRM communities. For example, there is a greater profit margin in processing pineapples into juice than in selling the pineapples themselves. Water for agriculture and industry can mean the difference between food staples and cash crops as well as between subsistence and adequate incomes.

There are strong links between water and energy that tend to manifest at larger scales, but advances in technology mean that energy production can occur at the local level e.g. through low head hydroelectric power generation. Energy can be a key factor in water security at the community level, particularly when groundwater forms a significant proportion of community water resources. Energy is required to pump the water out of the ground as well as for most forms of irrigation. Energy is also required for water treatment. Conversely, energy can be generated from human waste. Anaerobic digestion, the breaking down of waste by bacteria which live in oxygen-free environments, produces a gas and slurry by-product as a result of the process. The gas, which is 60% or more methane content, can be used for heating and lighting, or to generate electricity (Schuster-Wallace 2015). The slurry can be burned for fuel, or used as a fertilizer.

10.2.5 Water for the Environment

The water needs of the environment tend to be de-prioritized against other more compelling elements of water security. The environment, however, provides many services that contribute to water security, including water purification, a buffer against extreme events, food, and water storage. Environmental services tend to be more important for people living in RRM communities, and particularly for the poor, who cannot afford to purchase food that can be procured free of charge, and indigenous populations who traditionally live off the land. Wetlands support fish and plant-based food sources and can be key water sources for both people and their livestock. However, not all environmental services are available to local populations, with private property owners securing access to wetlands and lakes which fall on their property. Overall, it has been estimated that water-related ecosystem services should be valued in excess of $33 trillion per year (Costanza et al. 1997). More specifically, the services provided by wetlands, lakes and rivers have been valued at $1500 to $3000 per hectare per year (de Groot et al. 2012).

10.3 Achieving Community Water Security

While community water security has been demonstrated to have multiple dimensions, water security is not simply about securing sufficient quantities (and quality) of water for the different purposes articulated i.e. domestic use, food security, environmental security, and livelihoods. In many cases, it is assumed that putting infrastructure in place (i.e. technological solutions) is all that is required. Even if this was the case, the lack of recognition and inclusion of operation and maintenance requirements undermine sustainability (WHO 2014d). A significant body of evidence is developing, especially around sanitation scale-up, which suggests that technology in and of itself, is insufficient to bring about sustained water security, particularly at the community level. Indeed, at this local level, culture, knowledge, attitudes and practices, and perceptions, will either facilitate or inhibit change towards a more water secure community. However, without political will, even the best intentions may be undermined. Many barriers to change can also been seen as facilitators, depending on the individual perspective and on whether they are acting in a constrictive or empowering manner. For example, lack of capacity could describe an inability to achieve higher levels of education or training (which is a barrier) versus an ability but lack of opportunity to access additional education or training.

Perhaps the most important criterion for moving towards a water secure community is readiness (and willingness) to change. Resources can be sourced from many different places, but without a community ready to embrace and sustain change, investments will be short-lived, facilities not used, and therefore benefits not realized. However, motivation cannot achieve change without the resources (human, technical, financial, institutional, managerial, and political) to realize change. These resources are unlikely to be completely self-supplied, particularly in RRM contexts, which tend to be resource poor in a number of different areas, such as data, capacity, technology, electricity, accessibility, and land ownership. However, potential community resources should never be underestimated, particularly for economic, agricultural, and WaSH investments which accrue significant social, environmental, and economic benefits for community members. For example, small loans to community groups invested in any one of these sectors can be sound investments in creating and sustaining change (Mengueze et al. 2014), especially if supported through government programming and technical expertise.

Unfortunately, resources do not always exist where they are needed, when they are needed. There is a significant capacity deficit for delivering water security in RRM communities. Qualified personnel sourcing and retention can be difficult in these communities given location, small and distributed populations, and lack of services. RRM communities may be isolated from each other, but also from government resources. On the other hand, social capital and the sense of community can be much stronger in RRM communities than, for example, in urban settings.

Inappropriate technologies and the lack of consideration for operation and maintenance in addition to capital costs is another barrier to water security at this scale. Electric pumps with the capacity to over pump aquifers, dams to increase potable water without thought to wastewater treatment, new schools and healthcare facilities without adequate WaSH provisioning; all demonstrate technical failure. Indeed, technology diffusion and scale up will only occur for the right solution in the right place at the right time – a solution that fits local capacity and financing to construct, operate, and manage, and the available energy and supplies to continue operation. In addition to physical resources, sustainable diffusion and scale up requires institutional and political capacity and will to support local action.

10.4 Principles for Water Secure Communities

Building upon the need for water security at the community scale and the barriers and facilitators identified in the previous section, there are arguably five principles that must be adhered to in order to build water secure communities:

- Access to adequate resources
  While financing and technology are obviously required resources, time and expertise for mentorship and collaboration are equally important.

- Evidence-informed decision-making
  Even in low resource settings, traditional knowledge and baseline data can improve decision-making, prioritization and resource allocation.
- Knowledge sharing
  Bridging research, policy, and practice and engaging across stakeholders to share not only what works, but what did not, and why.
- Sustainability and innovation
  Understanding social, environmental and economic contexts to ensure appropriate solutions, building in operation, maintenance, and expansion, and innovating with what exists.
- Human rights
  A rights based approach should not be seen as abdication of responsibility on the part of individuals, but rather as progressive realization of equitable and inclusive government services that achieve these rights.

10.5 Measuring Water Security at the Community Level

As discussed to this point, water security is dependent on numerous social, political, technical, institutional, and environmental factors. Thus an holistic approach to measuring water security, incorporating all of these factors, is necessary to gain an accurate representation of the state of a community's water security. Only in this manner can true evidence-informed decision-making occur. However, in resource-poor contexts, data collection is not prioritized and many RRM communities do not have information readily available. Moreover, it is important to obtain both quantitative and qualitative data focusing on the unique attributes of any community, as these social and environmental factors can vary greatly between communities. For example, the number of people continuing to practice open defecation after pit latrines have been installed is a useful quantitative measure of the effectiveness of a pit-latrine intervention; however, these numbers do not provide information regarding why the intervention was or was not successful. Without this reason, it is difficult to know how to promote the use of pit latrines, particularly if the intervention was not successful or if the promotion is to be expanded to other communities. For example, collecting the qualitative data which provides the understanding that, when a person is suffering from diarrhoea it is much more difficult to squat over a pit latrine than it is to hang from a branch outside, is invaluable. Only with this information is it possible to know that improving sanitation facilities with hand rails could increase the number of people accessing pit latrines beyond current levels.

A number of tools are available in the literature in the shape of forms, checklists, and questionnaires designed to measure particular aspects of water security at the community level. Many of these tools focus on the drinking water supply chain, which is the path water follows from the source to consumption. It is important to note that none of these tools are holistic; they do not consider all dimensions of community water security shown in Fig. 10.1 and therefore they do not provide the entire picture of water security in a community. It is important to choose a tool, or suite of tools, that are capable of providing information relevant to the point at issue.

The WHO has developed a preventative management framework for safe drinking water incorporating several elements of water security. The framework incorporates (1) the establishment of health-based targets, including water quality targets, for drinking-water; (2) the development and implementation of water safety plans (WSPs) as a quality management system from source to consumption; and (3) a system of independent surveillance that verifies the WSPs are working. Many forms and checklists are compiled in the WSP manual (WHO and IWA 2009). The WSPs indicate that safe drinking water is achieved by protecting or treating water at all points along the drinking water supply chain including the source, transportation, distribution, storage, and handling (Davison et al. 2005). Gathering a set of reference data on the sanitary environment and hygienic handling of water along the drinking water supply chain to the point-of-use is a key aspect of developing a picture of a community's water security. Sanitary risk inspections are checklists that do exactly this; they provide a rapid assessment tool used to evaluate the likelihood of contamination based on the sanitary conditions at any point within the drinking water supply chain. The Rapid Assessment Drinking Water Quality (RADWQ) tool (WHO and UNICEF 2012) provides a methodology for collecting baseline information on drinking water quality required for target setting and independent surveillance.

From a qualitative perspective, a number of questionnaires have been developed to investigate the water-related local sociocultural factors such as the knowledge, attitudes and practices (KAP) posing a risk to water security - The W:ISE Toolkit (Schuster-Wallace et al. 2015a). The Community Self-Water Assessment Tool (Community SWAT) is a computer-based tool that combines information from these qualitative surveys with numerous forms, questionnaires, checklists, indices and indicators to support RRM communities in an holistic self-assessment of each dimension of their water security, including: water resources, water infrastructure, sanitation/wastewater infrastructure, access and equity, community capacity and capital, health and wellbeing, food security, economic activity, and environment. The aim of this assessment is (1) to facilitate a systematic consideration of a community's water security issues and (2) to consolidate the results into key graphics that could help the user identify relative strengths and threats (Schuster-Wallace et al. 2015a). This tool intends to serve communities by bringing awareness to the holistic nature of water security and by acting as a front-end for decision support as well as an evaluation of interventions to assess improvements across these dimensions. Incorporating Community SWAT with a tool that measures community wellbeing, such as the Canadian Index of Wellbeing (University of Waterloo 2011), provides evaluation of both direct and indirect benefits of enhanced community water security.

10.6 Integrated Pathways to Water Security

Building upon the principles articulated in the previous section, community water security requires an integrated coupled systems approach supported by institutional, political, and community capacity, the elements for which are shown in Fig. 10.2. While shared ownership of problems and solutions and multi-stakeholder processes require time and resources, sustainable development is mutual development, and water security must be embedded within the context of, and provide for, sustainable development. They are not mutually exclusive. Thus, water security at the community level requires two simultaneous pathways.

The first pathway is supportive capacity building, dialogue, and evidence for a process of community learning, visioning, and development that ultimately leads to

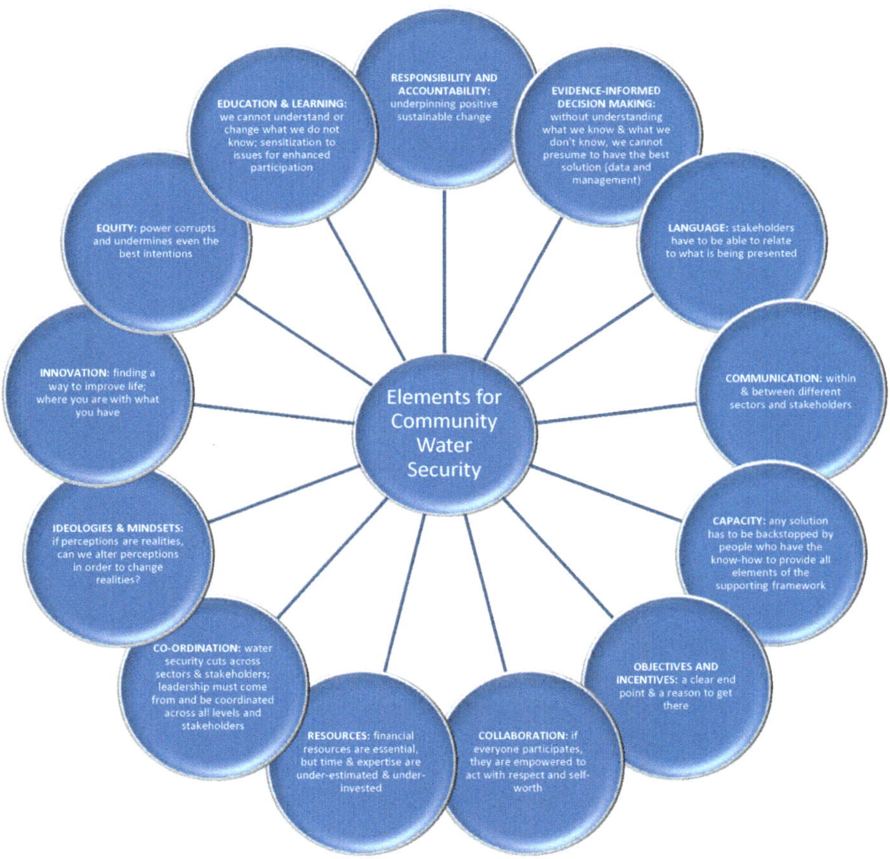

Fig. 10.2 Elements for achieving community water security (Modified from Schuster-Wallace et al. 2015b)

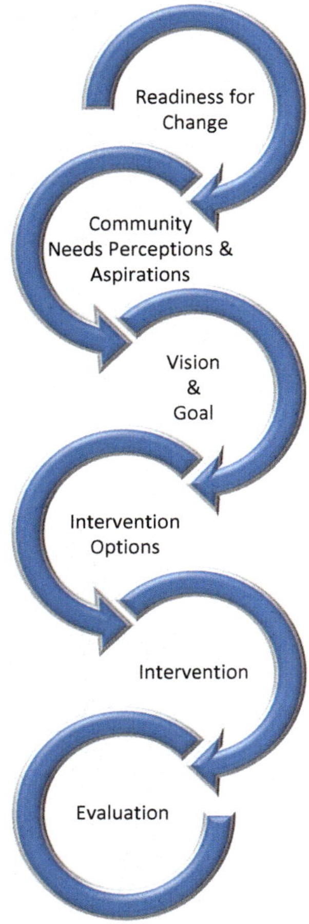

Fig. 10.3 Pathway for sustainable community change (From Schuster-Wallace et al. 2015b)

community derived, evidence informed decision making (Fig. 10.3) for identification of culturally sensitive and appropriate solutions (W:ISE Toolkit).

The second pathway is management of the resource itself for human, environmental, and economic security. This pathway must be embedded in an integrated, multiple use, systems approach to water resources management. It requires comprehensive consideration of water uses, demands, and risks for domestic water security, food security, environmental security, and livelihoods.

Traditional funding mechanisms must be augmented by innovative financing mechanisms to underpin these pathways. Innovative financing through activities which contribute to water security can include resource trading, such as water quality (see, for example, Conservation Technology Information Centre 2006) where a high polluter buys credits from a low polluter. Pollution reduction can be realized through conservation practices or other pollution-mitigation interventions, the cost of which is offset by revenue earned through trading. This mechanism has been used to reduce carbon emissions, albeit with challenges (see, for example Perdan and Azapagic 2011). More local innovations that can be used to finance WaSH, in particular, include microfinance and the Waste to Wealth approach. Microfinance coupled with community savings groups and reduced water fees for members could catalyze both improvements in community water and sanitation supply and social development (Mengueze et al. 2014). The Waste to Wealth approach, developed out of a Government of Canada supported Grand Challenges Canada grant, is a framework for realizing the value to be found in byproducts from human waste i.e. fertilizer and fuel. Anaerobic digestion fed by sanitation facilities and pumped sewage pits and tanks produces high grade fertilizer and biogas, a gas mixture with up to 70% methane content that can be used for cooking and lighting or generating electricity (Schuster-Wallace et al. 2014). In this manner money is made through activities which enhance community water security, simultaneously amplifying progress towards a water secure community. Key to success within this water security framework is the social reinvestment from innovative financing that supports community water management, social development, and community empowerment.

10.7 Conclusion

By defining community water security as part of the water security continuum with properties of sustainability, accessibility, affordability, and reliability as they pertain to both water quality and quantity, within and outside the community for health, dignity, and productive lives, we automatically include water in all its forms and for all its purposes. As such, it becomes apparent how community water security is nested within the larger continuum of multi-scale water security. Moreover, it becomes apparent that community water security is more than the individual uses of water in a community, more than competition for finite resources, and more than disaster risk reduction, although all of these are important to water security at the community level. Instead, community water security is the manifestation of a quasi-stable social-physical system in which supplies and demands are managed for sustainable use and maximum societal benefit. In this manner, a comprehensive systems approach to community-based water security coupled with capacity for change, particularly in the RRM context, is a viable way forward for sustainable development and the implementation of water-related SDG targets.

Acknowledgements Some of the ideas presented here have evolved out of the WaSH and Wellbeing: Changing Evidence, Evidence for Change Symposium (October 2013) funded by the Canadian Institutes for Health Research, the Water Institute at the University of Waterloo and the United Nations University Institute for Water, Environment and Health.

References

Asian Development Bank. (2013). *Asian water development outlook 2013*. Asian Development Bank and Asia-Pacific Water Forum. http://www.adb.org/publications/asian-water-development-outlook-2013. Accessed Jan 2016.

Bogardi, J. J., Dudgeon, D., Lawford, R., Flinkerbusch, E., Meyn, A., Pahl-Wostl, C., et al. (2012). Water security for a planet under pressure: Interconnected challenges of a changing world call for sustainable solutions. *Current Opinion in Environmental Sustainability, 4*, 1–9.

Bradley, J., & Bartram, J. K. (2013). Domestic water and sanitation as water security: monitoring, concepts and strategy. *Philosophical Transactions of the Royal Society A, 371*. doi:10.1098/rsta.2012.0420.

Calow, R. C., MacDonald, A. M., Nicol, A. L., & Robins, N. S. (2010). Ground water security and drought in Africa: Linking availability, access, and demand. *Ground Water, 48*(2), 246–256.

Clasen, T., Prüss Ustün, A., Mathers, C. D., Cumming, O., Cairncross, S., & Colford Jr., J. M. (2014). Estimating the impact of unsafe water, sanitation and hygiene on the global burden of disease: evolving and alternative methods. *Tropical Medicine & International Health, 19*(8), 884–893. doi:10.1111/tmi.12330.

Conservation Technology Information Centre. (2006). *Getting paid for stewardship: An agricultural community water quality trading guide*. http://ctic.org/media/users/lvollmer/pdf/GPfS_final(1).pdf. Accessed Jan 2016.

Cook, C., & Bakker, K. (2012). Water security: Debating an emerging paradigm. *Global Environmental Change, 22*, 94–102. doi:10.1016/j.gloenvcha.2011.10.011.

Cosgrove, W.J., & Rijsberman, F.R. (2000). *World water vision. Making water everybody's business*. Earthscan. http://www.worldwatercouncil.org/index.php?id=961. Accessed Jan 2016.

Costanza, R., d'Arge, R., de Groot, R., Farberk, S., Grasso, M., Hannon, B., et al. (1997). The value of the world's ecosystem services and natural capital. *Nature, 387*, 253–260.

Davison, A., Howard, G., Stevens, M., Callan, P., Fewtrell, L., Deere, D., & Bartram, J. (2005). *Water safety plans: managing drinking-water quality from catchment to consumer*. World Health Organization. WHO/SDE/WSH/05.06. http://www.who.int/water_sanitation_health/dwq/wsp170805.pdf. Accessed Jan 2016.

de Groot, R., Brander, L., van der Ploeg, S., Costanza, R., Bernard, F., Braat, L., et al. (2012). Global estimates of the value of ecosystems and their services in monetary units. *Ecosystem Services, 1*, 50–61.

Dickson, S. E., Schuster-Wallace, C. J., & Newton, J. J. (2016). Water security assessment indicators: The rural context. *Water Resources Management, 30*(5), 1567–1604.

El Ganzouri, A., Schuster-Wallace, C.J., & Dickson S.E. (2015). Rainwater harvesting: a necessary but insufficient solution to meeting the human right to drinking water in Uganda. In A. El Ganzouri (Ed.), *Modeling and analysis of rainwater harvesting systems under different climates*. M.A.S. Thesis. http://hdl.handle.net/11375/16445

Falkenmark, M. (2001). The greatest water problem: The inability to link environmental security, water security and food security. *Water Resources Development, 17*(4), 539–554.

Grey, D., & Sadoff, C. W. (2007). Sink or swim? Water security for growth and development. *Water Policy, 9*, 545–571.

Grey, D., Garrick, D., Blackmore, D., Kelman, J., Muller, M., & Sadoff, C. (2013). Water security in one blue planet: Twenty-first century policy challenges for science. *Philosophical Transactions of the Royal Society A, 371*, 20120406. doi:10.1098/rsta.2012.0406.

Hutton, G. (2012). Global costs and benefits of drinking-water supply and sanitation interventions to reach the MDG target and universal coverage WHO/HSE/WSH/12.01. http://www.who.int/water_sanitation_health/publications/2012/globalcosts.pdf. Accessed Jan 2016.

Jury, W. A., & Vaux Jr., H. J. (2007). The emerging global water crisis: Managing scarcity and conflict between water users. *Advances in Agronomy, 95*, 1–76.

Lundqvist, J., Appasamy, P., & Nelliyat, P. (2003). Dimensions and approaches for Third World city water security. *Philosophical Transactions of the Royal Society of London B, 358*, 1985–1996. doi:10.1098/rstb.2003.1382.

Mengueze, S., Mbuvi, D., Dickin, S., & Schuster-Wallace, C.J. (2014). *A micro financing framework for rural water and sanitation provisioning in Sub-Saharan Africa*. UNU INWEH and UNU MERIT. http://inweh.unu.edu/wp-content/uploads/2014/08/A-Mircofinancing-Framework-Report.pdf. Accessed Jan 2016.

Milly, P. C. D., Betancourt, J., Falkenmark, M., Hirsch, R. M., Kundzewicz, Z. M., Lettenmaier, D. P., & Stouffer, R. J. (2008). Stationarity is dead: Whither water management? *Science, 319*, 573–574.

Molle, F., & Mollinga, P. (2003). Water poverty indicators: Conceptual problems and policy issues. *Water Policy, 5*, 529–544.

Mullin, P. M., & Incerpi, M. H. (2010). Fetal growth restriction. In T. M. Goodwin, M. N. Montoro, L. Muderspach, R. Paulson, S. Roy, et al. (Eds.), *Management of common problems in obstetrics and gynecology* (pp. 21–25). Chichester: Wiley.

OECD. (2013). *Water security for better lives*. OECD Publishing. http://www.oecd.org/env/resources/watersecurity.htm Accessed Jan 2016.

Onda, K., LoBuglio, J., & Bartram, J. (2012). Global access to safe water: Accounting for water quality and the resulting impact on MDG progress. *International Journal of Environmental Research and Public Health, 9*(3), 880–894. doi:10.3390/ijerph9030880.

Perdan, S., & Azapagic, A. (2011). Carbon trading: Current schemes and future developments. *Energy Policy, 39*(10), 6040–6054 http://dx.doi.org/10.1016/j.enpol.2011.07.003.

Prendergast, A., & Kelly, P. (2012). Enteropathies in the developing world: Neglected effects on global health. *The American Journal of Tropical Medicine and Hygiene, 86*(5), 756–763. doi:10.4269/ajtmh.2012.11-0743.

Prüss Ustün, A., Bos, R., Gore, F., & Bartram, J.. (2008). Safer water, better health: Costs, benefits and sustainability of interventions to protect and promote health. World Health Organization, Geneva. http://whqlibdoc.who.int/publications/2008/9789241596435_eng.pdf. Accessed Jan 2016.

Roy, S. (1999). De-development revisited: Palestinian economy and society since Oslo. *Journal of Palestine Studies, 28*(3), 64–82.

Schuster-Wallace, C. J., & Watt, M. S. (2015). Women and the water-health nexus. In J. Chamberlain Froese & L. Elit (Eds.), *Women's health in the majority world* (2nd ed.). Hauppauge: Nova Sciences Publishers.

Schuster-Wallace, C.J., Cave, K., Metcalfe, C., Theodoulou, M., & Yargeau, V. (2014). *From waste to wealth: Sustainable wastewater management in Uganda wastewater: Management framework*. UNU INWEH. http://inweh.unu.edu/wp-content/uploads/2014/06/Wastewater-Management-Framework_FINAL.pdf. Accessed Jan 2016.

Schuster-Wallace, C. J. (2015). Waste to wealth: Helping to close the sanitation financing gap in rural communities and small towns. *The Solutions Journal, 6*(6), 60–69.

Schuster-Wallace, C.J., Watt, M.S., Cave, K., Dickson, S.E., et al. (2015a). *The W:ISE handbook: An integrated social empowerment toolkit for community wash and wellbeing*. UNU INWEH. http://inweh.unu.edu/wp-content/uploads/2016/04/WISE-Toolkit-Manual_web.pdf. Accessed Jan 2016.

Schuster-Wallace, C.J., Watt, M.S., Cave, K., & Mehta, P. (2015b). *Doing development differently: Bridging policy, practice, and research for WaSH and wellbeing*. UNU INWEH. http://inweh.unu.edu/wp-content/uploads/2016/04/Mucky-Middle-report_April25_reduced.pdf. Accessed Jan 2016.

Siwar, C., & Ahmed, F. (2014). Concepts, dimensions and elements of water security. *Pakistan Journal of Nutrition, 13*(5), 281–286.

Smith, L., & Haddad, L. (2014). *Reducing child undernutrition: Past drivers and priorities for the post-MDG Era*. IDS Working Paper 441 http://opendocs.ids.ac.uk/opendocs/bitstream/handle/123456789/3816/Wp441R.pdf?sequence=4. Accessed Jan 2016.

UNICEF. (2014). *WASH package for Ebola care and treatment centres/units. Guidance note.* UNICEF Programme Division. http://reliefweb.int/report/liberia/wash-package-ebola-care-and-treatment-centresunits-guidance-note. Accessed Jan 2016.

University of Waterloo. (2011). Canadian Index of Wellbeing. University of Waterloo. https://uwaterloo.ca/canadian-index-wellbeing/. Accessed Jan 2015.

UNESCO. (2009). *Water in a changing world.* World Water Development Report 3. UN Water. http://unesdoc.unesco.org/images/0018/001819/181993e.pdf. Accessed Jan 2016.

WHO. (1986). Ottawa Charter for Health Promotion. World Health Organization. http://www.who.int/healthpromotion/conferences/previous/ottawa/en/. Accessed Jan 2016.

WHO. (2014a). World health day: Vector borne diseases http://www.who.int/campaigns/world-health-day/2014/vector-borne-diseases/en/. Accessed Jan 2016.

WHO. (2014b). Dengue and severe dengue. WHO Fact sheet N°117. Updated March 2014. http://www.who.int/mediacentre/factsheets/fs117/en/. Accessed Jan 2016.

WHO. (2014c). World Malaria Report 2014. WHO. http://apps.who.int/iris/bitstream/10665/144852/2/9789241564830_eng.pdf. Accessed Jan 2016.

WHO. (2014d). *Investing in water and sanitation: Increasing access, reducing inequalities.* UN-Water global analysis and assessment of sanitation and drinking-water (GLAAS) 2014 report. UN Water. http://www.who.int/water_sanitation_health/publications/glaas_report_2014/en/. Accessed Jan 2016.

WHO and IWA. (2009). *Water safety plan manual* (WSP Manual). Step-by-step risk management for drinking water supplies. http://apps.who.int/iris/bitstream/10665/75141/1/9789241562638_eng.pdf. Accessed Jan 2016.

WHO and UNICEF. (2012). *Rapid assessment of drinking-water quality: A handbook for implementation.* Joint Monitoring Programme. http://www.who.int/water_sanitation_health/publications/2012/rapid_assessment/en/. Accessed Jan 2016.

WHO and UNICEF. (2015). *Progress on sanitation and drinking water - 2015 update and MDG assessment.* Joint Monitoring Programme. http://www.wssinfo.org/fileadmin/user_upload/resources/JMP-Update-report-2015_English.pdf. Accessed Jan 2016.

World Bank. (2008). *Agriculture for development.* World Development Report. World Bank. http://siteresources.worldbank.org/INTWDR2008/Resources/WDR_00_book.pdf. Accessed Jan 2016.

Index

D. Devlaeminck et al. (eds.), *The Human Face of Water Security*, Water Security in a New World, DOI 10.1007/978-3-319-50161-1

Y

Z

Printed by Printforce, the Netherlands